T0252344

ANNALS OF MATHEMATICS STUDIES
NUMBER 10

TOPICS IN TOPOLOGY

BY

SOLOMON LEFSCHETZ

PRINCETON
PRINCETON UNIVERSITY PRESS
LONDON: HUMPHREY MILFORD
OXFORD UNIVERSITY PRESS
1942

Lithoprinted in U.S.A.
EDWARDS BROTHERS, INC.
ANN ARBOR, MICHIGAN
1942

INTRODUCTION

The present monograph has been planned in such a way as to form a natural companion to the author's volume Algebraic Topology appearing at the same time in the Colloquium Series and hereafter referred to as AT. The topics dealt with have for common denominator the relations between polytopes and general topology. The first chapter takes up the relations between polytopes in general and the topologies which they may receive and in these questions we lean particularly heavily upon J. Tukey. The second chapter completes in certain important points the treatment of singular elements of AT. The third chapter deals with mappings of spaces on polytopes and certain related imbedding questions; it contains also a modern treatment of retraction for separable metric spaces. The last chapter is devoted to the group of questions centering around the general concept of local connectedness. Comparisons with retracts are considered at length, there is a full treatment of the homology and fixed point properties. The chapter concludes with an outline of the relations with "homology" local connectedness (the so-called HLC properties).

The general notations are those of AT. In addition to a short reference bibliography, a mere supplement to that of AT, there is also given a fairly comprehensive bibliography on locally connected spaces and retraction.

TABLE OF CONTENTS

Chapter I.

POLYTOPES

§1. AFFINE SIMPLEXES AND COMPLEXES

1. Affine Simplexes. In spite of the evident analogy with the treatment of Euclidean simplexes of (AT, III, VIII). it will be more convenient and also clearer to repeat the necessary introductory definitions and properties.

Our simplexes are considered here also as subsets of a real vector space $\mathfrak{B}$ whose elements are to be called points.

(1.1) DEFINITION. Let $\sigma^p = a_0 \dots a_p$ be a p-simplex whose vertices are independent points of a real vector space $\mathfrak{B}$. By the affine p-simplex associated with σ^p is meant the set, written σ^p_v given by

(1.2) $$x = x^i a_i$$

(1.3) $$p = 0 : x^0 = 1,$$

(1.4) $$p > 0 : 0 < x^i < 1, \quad \sum x^i = 1.$$

The x^i's are the barycentric coordinates of x. To the face $\sigma^q = a_{i_0} \dots a_{i_q}$ of σ^p there corresponds the set of points obtained by replacing $0 < x_{i_h}$ by $0 = x_{i_h}$ in (1.4); it is the σ^q_v associated with σ^q and is called a q-face of σ^p_v. We transfer to σ^p_v and to its faces the terminology previously adopted for σ^p. In particular we speak of the open or closed affine simplex, the <u>boundary</u> $\mathfrak{B}\sigma_v^p$ etc. The set of all points in an element of $\mathfrak{B}\sigma_v^p$ or of $\mathrm{Cl}\,\sigma_v^p$ is denoted by $|\mathfrak{B}\sigma_v^p|$ or $|\mathrm{Cl}\,\sigma_v^p|$.

(1.5) The open and the closed affine simplexes are convex.

Let $x', x'' \in \mathrm{Cl}\,\sigma_v^p$. The segment $l = \overline{x'x''}$ joining them consists of the points

(1.6) $x = t'x' + t''x'', \; 0 \leq t', t'' \leq 1, t' + t'' = 1.$

Hence if $x' = x^{i}a_i$, $x'' = x''^{i}a_i$ we have

$$x = x^{i}a_i, \; y^i = t'\,x'^{i} + t''\,x''^{i}$$

and we verify readily that $x \in |\mathrm{Cl}\,\sigma_v^p|$. Similarly for σ_v^p.

(1.7) If $\sigma_v^p = \sigma_v'\sigma_v''$ (complementary faces) there passes through each point x a unique segment $\overline{x'x''}$ with $x' \in \sigma_\alpha'$, $x'' \in \sigma_\alpha''$.

(Same proof as for (AT,VIII,2.1).

2.(2.1) DEFINITION. Let $S = \{\sigma_{vi}\}$, $S' = \{\sigma_{vi}'\}$ be two sets of affine simplexes, where the simplexes in each set are disjoint. We shall say that S' is a <u>simplicial</u> partition of S whenever each σ_{vi}' is in some σ_{vj} and each σ_{vj} is a union of a finite number of σ_{vi}'. Thus S' is a partition of S in the sense of (AT,IV,29).

(2.2) Let $S = \{\sigma_{vi}\}$ be a simplicial partition of $\mathfrak{B}\sigma_v^p$ and $\hat{\sigma}^p$ any point of $\bar{\sigma}_v^p$. Then: (a) if $\hat{\sigma}^p \subset \sigma_v^p$, $S' = \{\hat{\sigma}^p, \; \hat{\sigma}^p\sigma_{vi}\}$ is a simplicial partition of σ_v^p; (b) if $\hat{\sigma}^p \in \sigma_{vi}$, $S' = \{\hat{\sigma}^p\sigma_{vj} \mid j \neq i\}$ has the same property.

Since (2.2) is trivial for $p = 0$ we assume $p > 0$. Suppose first $\hat{\sigma}^p \subset \sigma^p_v$ and let $x \neq \hat{\sigma}^p$. By (1.5) the segment $\hat{\sigma}^p x$ extended meets $|\mathfrak{B}\sigma^p_v|$ in a point x' in some σ_{vi} and so $x \in \hat{\sigma}^p \sigma_{vi}$. Thus σ^p_v is the union of the elements of S'. Since $\hat{\sigma}^p$ is in no $\hat{\sigma}^p \sigma_{vi}$ we only have to prove the disjunction property for a pair $\hat{\sigma}^p \sigma_{vi}, \hat{\sigma}^p \sigma_{vh}$, $i \neq h$. Now if x is a point common to both, $\hat{\sigma}^p x$ extended will meet $\mathfrak{B}\sigma^p_v$ in a point common to σ_{vi}, σ_{vh} and this is ruled out since S is a simplicial partition of $\mathfrak{B}\sigma^p_v$. The treatment of (b) is essentially similar.

(2.3) Let $\{\sigma_{vi}\}$ be the set of all the proper faces of σ^p_v and $\hat{\sigma}_i, \hat{\sigma}^p$ points on σ_{vi}, σ^p_v. Then the affine simplexes

(2.4) $\zeta = \hat{\sigma}_1 \ldots \hat{\sigma}_j \hat{\sigma}^p, \ \sigma_{vi} \prec \cdots \prec \sigma_{vj}$

make up a simplicial partition of σ^p_v.

This is trivial for $p = 0$ so we assume it for dimensions $< p$ and prove it for p. Under the hypothesis of the induction the collection of all the $\zeta'_0 = \hat{\sigma}_1 \ldots \hat{\sigma}_j$ $\sigma_{vi} \prec \cdots \prec \sigma_{vj}$ terminating with $\hat{\sigma}_j$ is a simplicial partition of σ_{vj}. Since the σ_{vj} are disjoint $\{\zeta'\}$ is a simplicial partition of $\mathfrak{B}\sigma^p_v$, so that (2.3) follows now from (2.2).

The decomposition of $(\mathrm{Cl}\ \sigma^p_v)$ by the simplexes (2.11) is its first derived $(\mathrm{Cl}\ \sigma^p_v)'$. Usually the centroid $(\frac{1}{p+1}, \ldots, \frac{1}{p+1})$ is chosen as $\hat{\sigma}^p$ and similarly for the faces. The corresponding $(\mathrm{Cl}\ \sigma^p_v)'$ is known as the barycentric first derived. We can treat similarly the simplexes of $(\mathrm{Cl}\ \sigma^p_v)'$, and obtain the successive derived or barycentric derived as the case may be. In general, unless otherwise stated, "derived" shall stand for "barycentric derived".

(24.5) The following designations will be found very convenient. The simplexes of $(\mathrm{Cl}\ \sigma^p_v)^{(n)}$ will be designated by σ_n (we omit the subscript v). Since the σ_n

make up a dissection of $\mathrm{Cl}\,\sigma_v^p$ every point x of the latter belongs to one and only one σ_n which will be denoted by $\sigma_n(x)$.

3. The vector space $\mathfrak{V}$ or its subspaces may be metrized in various ways. For our purpose it is sufficient to consider an Euclidean metric relative to a base $B = \{b_i\}$. If $x = x^i b_i$, $y = y^i b_i$ (finite sums) such a metric is defined by

$$d(x,y) = \left(\sum (x^i - y^i)^2 \right)^{1/2} \tag{3.1}$$

and it has a meaning for all (x,y). The simplexes of $\mathfrak{V}$ are then Euclidean and may be written σ_e^p as in AT. The simplexes of the nth derived $(\mathrm{Cl}\sigma_e^p)^{(n)}$ have a maximum diameter: the mesh of the derived.

As a special case one may utilize the metric (3.1) attached to the subspace spanned by the vertices a_i of σ^p in $\mathfrak{V}$ relative to the base $\{a_i\}$ for the subspace. We thus obtain a metric for σ^p, and in fact for $|\mathrm{Cl}\sigma_v^p|$, given by (3.1) where x^i, y^i are now the barycentric coordinates of x, y. This particular metric will be called the natural metric of σ_v^p. Notice that if $\sigma_v^q \prec \sigma_v^p$, the induced metric in σ_v^q, is likewise its natural metric.

(3.1a) Remark. If no topology is specified for σ_v^p it will be understood that the set has been topologized by means of its natural metric. In point of fact the various topologies that may be specified in the sequel for σ_v^p will always be equivalent to the one induced by its general metric. This property is readily verified in each case and no further reference will be made to it later.

(3.2) The Euclidean p-simplex σ_e^p is a p-cell; its boundary $\mathfrak{B}\sigma_e^p$ is a (p-1)-sphere and $\bar{\sigma}_e^p$ is a p-dimensional parallelotope.

This is a consequence of (AT, I, 12.9) and the fact that $|\mathrm{Cl}\,\sigma_v^p|$ is a bounded convex subset of a Euclidean space $\mathfrak{E}^p$, the metrized subspace of the vertices.

(3.3) Let $\sigma_e^p = a_0 \ldots a_p$, be a simplex in an Euclidean space $\mathfrak{E}^n$ and x any point of $\mathfrak{E}^n$. Then $d(x,y)$, $y \in \sigma_e^p$, does not exceed the maximum distance ρ from x to the vertices. (AT, VIII, 2.2).

(3.4) The diameter of σ_e^p is the length of its longest edge (AT, VIII, 2.3).

(3.5) Mesh $(\mathrm{Cl}\,\sigma_e^p)' \leqq \frac{p}{p+1}$ diam σ_e^p. (AT, VIII, 2.4).

(3.6) If σ_v^p is assigned the natural metric then mesh $(\mathrm{Cl}\,\sigma_v^p)^{(n)} \leqq \sqrt{2}\,(\frac{p}{p+1})^n$, which $\to 0$ as $n \to \infty$.

For in the natural metric the edges of σ_v^p all have the length $\sqrt{2}$ and so (3.6) is a consequence of (3.5).

(3.7) If x, x' are distinct points of σ_v^p there is an n such that the simplexes $\sigma_n(x)$, $\sigma_n(x')$ containing x, x' have no common vertices. (3.6).

(3.8) Let $\{\sigma_n\}$ be such that $\bar{\sigma}_{n+1} \subset \sigma_n$ (notations of 2.5). Then $\bigcap \bar{\sigma}_n = x$ a point of $\bar{\sigma}_v^p$.

In the natural metric $\bar{\sigma}_v^p$ is a compactum and $\{\bar{\sigma}_n\}$ a collection of closed subsets with the finite intersection property. Hence $\bigcap \bar{\sigma}_n \neq \phi$ and since diam $\bar{\sigma}_n \to 0$ the intersection is a point.

(3.9) Let $x \in \sigma_v^q \prec \sigma_v^p$. Then there exists an n such that $\sigma_n(x)$ has all its vertices in St σ_v^q (star in $\mathrm{Cl}\,\sigma_v^p$).

For diam $\sigma_n(x) \longrightarrow 0$ and the distance from x to the set of simplexes not in St σ_v^q is positive.

4. Affine complexes. Just as for simplexes it is convenient as well as clearer to separate the affine and other complexes. The affine complex serves to specify the point-set which under suitable topologies becomes a geometric or an Euclidean complex.

(4.1) DEFINITION. Let $K = \{\sigma\}$ be a simplicial complex and let $\{A_i\}$ be its vertices where $\{i\}$ is any set whatever. Let $\{a_i\}$ be vectors of a real vector space with the following properties:

(4.2) $a_i \longleftrightarrow A_i$ is one-one;

(4.3) if $\sigma = A_i \ldots A_j \in K$ then $a_i, \ldots, a_j$ are independent, and so they are the vertices of an affine simplex denoted by σ_v;

(4.4) $\sigma \not\supset \sigma' \Longrightarrow \sigma_v \cap \sigma'_v = \phi$.

If we transfer to $\{\sigma_v\}$ the incidences "is a face of" prevailing in K, likewise the same incidence-numbers, it becomes a complex $K_v \cong K$, known as an affine simplicial complex. Its relation to K is also described by the statement: K_v is an affine realization of K. We also refer sometimes to K as an antecedent of K_v.

We transfer to K_v the full terminology attached to K.

Example. $\mathrm{Cl}\,\sigma_v^p, \mathfrak{B}\sigma_v^p$ are affine realizations of $\mathrm{Cl}\,\sigma^p$, $\mathfrak{B}\sigma^p$ and σ_v^p is an open subcomplex of $\mathrm{Cl}\,\sigma_v^p$.

The set of all the points of the simplexes of K_v is denoted by $|K_v|$.

It follows from the definition of K_v that every point $x \in |K_v|$ satisfies a relation

(4.5) $$x = x^i a_i$$

where if $x \in \sigma_v$ considered in (4.3), the coordinates $x^i, \ldots, x^j$ are the barycentric coordinates of x in σ_v.

It follows that the x^i are unique and satisfy (1.3), (1.4). The x^i are known here also as the <u>barycentric coordinates</u> of x.

(4.6) <u>Barycentric mapping</u>. The definition is the same as for Euclidean complexes (AT,VIII,6.1) and need not be repeated.

(4.7) A noteworthy special case is when K_v, K_{1v} are both realizations of the same complex K. Let $\{A_i\}$, $\{a_i\}$, $\{a_i'\}$ be the vertices of K, K_v, K_{1v} where a_i, a_i' are the images of A_i. Then $a_i \longrightarrow a_i'$ is a one-one transformation which induces a one-one barycentric mapping τ, referred to as the <u>natural</u> barycentric mapping $K_v \longrightarrow K_{1v}$.

We notice the following properties:

(4.8) Every simplicial complex K has an affine realization K_v.

For if $\{A_i\}$ is chosen as a base for a real vector-space $\mathfrak{V}$ (its elements being all the finite forms t^iA_i with the t_i real) the three conditions (4.2), (4.3), (4.4) are naturally satisfied and so K_v may be constructed with $a_i = A_i$ throughout.

It is important to observe that this special choice of the a_i is not unique. Thus consider the two-complex K^2 consisting of a $\mathfrak{B}\sigma^3$ with one two-face removed. K^2 has the following affine realization: take a plane triangle ABC and let D be its centroid; K^2 consists of the triangles DAB, DBC, DCA with all their sides and vertices. This is a realization as a subset of a plane, whereas the above construction would require a four-space.

(4.9) Let $\hat{\sigma}$ be some point on $\sigma_v \in K_v$. Then:

(a) $\zeta = \hat{\sigma}_i \ldots \hat{\sigma}_j, \sigma_{vi} \prec \cdots \prec \sigma_{vj}$ is an affine simplex and

$$\zeta \subset \sigma_{vj};$$

(b) $K_v' = \{\zeta\}$ is an affine realization

of K', and is known as a first derived of K_v;

(c) $|K_v| = |K'_v|$.

This is an immediate consequence of (1.10) together with (AT,IV,26).

When the new vertices $\hat{\sigma}$ are the centroids of the corresponding σ, the affine complex K'_v is called the <u>barycentric first derived</u>. The definition of the nth derived, barycentric or otherwise is now obvious. It is written $K_v^{(n)}$ and is an affine realization of $K^{(n)}$ which coincides with K_v as a point set.

(4.10) <u>Notations</u>. Extending the notations introduced in (2.5) we designate by σ_n the simplexes of $K^{(n)}$ (also σ for σ_o) and by $\sigma_n(x)$ the $\sigma_n \ni x$.

The following property is needed later.

(4.11) Let $\hat{\sigma}$ be a point of σ_v and let K_v undergo the set-transformation S (in the sense of AT,IV,7): S is the identity outside of St σ_v; $S\sigma_v = \hat{\sigma}\mathfrak{B}\sigma_v$; if $\sigma'_v \in \mathrm{St}\,\sigma_v - \sigma_v$; $S\sigma'_v = \hat{\sigma}(\mathfrak{B}\sigma'_v - \sigma_v)$. Then S is a simplicial partition of K_v into a new complex K_{1v}, and K_{1v} is a subdivision of K_v.

The partition property is an immediate consequence of (1.9). It is also clear that S fulfills the conditions of (AT,IV,24.8) and so it is a subdivision.

(4.12) Consider the function $d(x,y)$ defined on K_v by the expression (3.1). If K_{1v} is the affine realization of (4.8) with $\{A_1\}$ as the base for the vector space $\mathfrak{V}$, then $d(x,y)$ is a metric for $\mathfrak{V}$ and hence for K_{1v}. Since the natural barycentric transformation $K_{1v} \to K_v$ is one-one and preserves the barycentric coordinates, d(x,y) is likewise a distance-function for K_v. We will call this metric <u>natural</u>. Evidently

> (4.13) All the affine realizations of the same K with their natural metric are topologically equivalent. More precisely their natural barycentric mappings into one another are topological and in fact isometric (distance-preserving).

§2. GEOMETRIC COMPLEXES

5.(5.1) The natural metric provides one mode of topologizing the set $|K_v|$. Another consists in assigning to $|K_v|$ as subbase for the open sets the totality of the stars of the vertices in all the derived $K_v^{(n)}$. When this topology is chosen the complex is said to be <u>geometric</u>, often denoted by K_g. The space $|K_v|$ with the above topology is written $|K_g|$ and called a <u>polytope</u>. The simplexes σ_r are also written accordingly σ_g and called <u>geometric simplexes</u>. As we shall see later (II, 5.1) the topology just chosen is the one required for our basic mapping theorem. We call K_v: an <u>affine antecedent</u> of K_g, and call a simplicial antecedent of K_v: a <u>simplicial antecedent</u> of K_g.

> (5.2) THEOREM. A polytope $|K_g|$ is metrizable, and hence it is a normal Hausdorff space (W. Wilson [a]; proof by J. Tukey).

In another form also: the open set topology just assigned to K_v to turn into a polytope is equivalent to a topology obtainable from a distance function.

We shall denote by $K^{(n)}$ the nth barycentric derived of K_g, by σ_n, $\sigma_n(x)$ the same as in (4.10) relative to $K_g^{(n)}$, by $\hat{\sigma}_n$ the vertex of $K_g^{(n+1)}$ in σ_n, or which is the same the centroid of σ_n. The distance is introduced as a limit of functions which satisfy the triangle axiom. In a complex a measure of the distance of two points is the length of the shortest chain of simplexes each incident with the next which joins the simplexes containing the points. This lies at the root of the metric.

6. Let a,b be any two vertices of K_g and let $f_o(a,b)$ denote the least n if any exists such that there is a finite sequence or "chain" of vertices $a = a_o, a_1, \ldots, a_n = b$, where $a_i a_{i+1}$ is a one-simplex of K_g. If no such chain exists we set $f_o(a,b) = \infty$; this last circumstance occurs when and only when a,b are in distinct components of K_g. A similar function may be introduced for $K_g^{(n)}$ and it is denoted by $f_n(a,b)$.

(6.1) If (a,b) are vertices of K_g and hence also of K_g' then $f_1(a,b) = 2f_o(a,b)$.

If $a = a_o, \ldots, a_n = b$ is as before then $a = a_o$, $\widehat{a_o a_1}, a_1 \ldots, a_n = b$ is a chain joining a to b in K_g' and so $f_1(a,b) \leq 2 f_o(a,b)$. If $a = \hat{\sigma}_o, \hat{\sigma}_1, \ldots, \hat{\sigma}_m = b$ is a shortest chain in K_g' then we must have $\sigma_o \prec \sigma_1 \succ \sigma_2 \ldots$, with alternating incidences. For if $\sigma_{i-1} \prec \sigma_i \prec \sigma_{i+1}$ or $\sigma_{i-1} \succ \sigma_i \succ \sigma_{i+1}$, then $\hat{\sigma}_i$ could be omitted. Hence we may replace $\hat{\sigma}_2, \hat{\sigma}_4, \ldots$ by vertices $a_1, a_2, \ldots$ and so $a = \sigma_o, a_1 a_2, \ldots, b$ is a chain from a to b in K_g. Therefore $m = f_1(a,b) \geq 2n$, and (6.1) follows.

We now extend f_o to $|K_g|$ by

$$f_o(x,y) = \inf \{f_o(a,b)\} + \begin{cases} 0 & \text{if } x,y \text{ are both vertices;} \\ \frac{1}{2} & \text{if only one of } x,y \text{ is a vertex;} \\ 1 & \text{if } x,y \text{ are not vertices,} \end{cases}$$

where a,b range respectively over the vertices of $\sigma(x)$, $\sigma(y)$. A similar function $f_n(x,y)$ is defined for $K_g^{(n)}$.

(6.2) Setting $d_n(x,y) = 2^{-n} f_n(x,y)$ we find readily:

$$0 \leq d_n(x,y); \quad d_n(x,y) = d_n(y,x);$$

$$d_{n+1}(x,y) \leq d_n(x,y); \quad d_n(x,z) \leq d_n(x,y) + d_n(y,z).$$

Hence the function $d(x,y) = \inf \{1, \lim d_n(x,y)\}$ clearly satisfies all but the last of the distance conditions (AT,I,42):

(6.3a) $0 \leq d(x,y)$,

(6.3b) $d(x,y) = d(y,x)$,

(6.3c) $d(x,z) \leq d(x,y) + d(y,z)$,

(6.3d) $d(x,y) = 0 \Longrightarrow x = y$,

and so we must verify (6.3d).

We first prove:

(6.4) If $d(x,y) \leq 2^{-(n+1)}$ then $\sigma_{n-1}(x)$ and $\sigma_{n-1}(y)$ have a common vertex.

Let a,b be vertices of $\sigma_n(x)$, $\sigma_n(y)$. We have:

$$d_n(a,b) = d_{n+m}(a,b) \leq d_{n+m}(a,x) + d_{n+m}(x,y) + d_{n+m}(y,b)$$
$$\leq d_n(a,x) + d_{n+m}(x,y) + d_n(y,b) \leq 3.2^{-(n+1)}.$$

Therefore $f_n(a,b) \leq \frac{3}{2}$, and hence $f_n(a,b) = 0$ or 1. It follows that if $a = \hat{\sigma}'_{n-1}$, $b = \hat{\sigma}''_{n-1}$ then $\hat{\sigma}'_{n-1} = \hat{\sigma}''_{n-1}$ or else $\hat{\sigma}'_{n-1}\,\hat{\sigma}''_{n-1}$ is a one-simplex of $K_g^{(n)}$. Consequently σ'_{n-1}, σ''_{n-1} are incident, and we will suppose say $\sigma'_{n-1} \prec \sigma''_{n-1}$.

Since $\hat{\sigma}'_{n-1}$ is a vertex of $\sigma_n(x)$ and $\bar{\sigma}_n(x) \subset \bar{\sigma}_{n-1}(x)$, we have $\sigma'_{n-1} \prec \sigma_{n-1}(x)$, and similarly $\sigma''_{n-1} \prec \sigma_{n-1}(y)$. Hence $\sigma'_{n-1} \prec \sigma_{n-1}(y)$ and so $\sigma_{n-1}(x)$ and $\sigma_{n-1}(y)$ have a common face and therefore a common vertex.

Suppose $d(x,y) = 0$. By (6.4), $\sigma_n(x)$ and $\sigma_n(y)$ have a common vertex for every n, and so by (3.7): $x = y$, which is (6.3d).

We notice now that $\{\mathrm{St}\ \sigma_n(x)\}$ is a base at the point x. In the metric just obtained $\operatorname{diam} \sigma_n(x) \leq 2^{-n} \longrightarrow 0$, and so every sphere $\mathfrak{S}(x, \rho)$ contains an $\mathrm{St}\sigma_n(x)$. On the other hand by (3.9) there is an m such that if

σ'_n is not in St $\sigma_n(x)$ then $\sigma_{n+m}(x)$ has no vertices in σ'_n. Therefore if $y \in \sigma'_n$ then $\sigma_{n+m}(x)$ and $\sigma_{n+m}(y)$ have no common vertices and hence by (6.4): $d(x,y) > 2^{-(n+m+3)}$. Consequently $G(x, 2^{-(m+n+3)}) \subset \mathrm{St}\ \sigma_n(x)$. Therefore the topology induced by $d(x,y)$ agrees with the topology of K_g, and the metrization theorem (5.2) is proved.

7. Euclidean complexes. We recall from (AT,III,6.9):

(7.1) DEFINITION. An Euclidean complex is a countable locally finite affine complex K_e situated in an Euclidean or Hilbert parallelotope and such that its elements $\{\sigma_e\}$, which are Euclidean simplexes, satisfy the condition

(7.2) $\sigma_e \cap \overline{|(K_e - \mathrm{St}\ \sigma_e)|} = \emptyset$.

If K_e is an affine realization of K we speak of it as an Euclidean realization of K.

(7.3) The derived complexes of an Euclidean complex are Euclidean.

It is sufficient to prove K'_e Euclidean. If we refer to the expression (2.4) for the simplexes of K'_e in terms of those of K_e we find first that each σ_e contains at most a finite number of simplexes σ_1 of K'_e, and so K'_e is countable. Furthermore if $\sigma_1 \prec \sigma'_1$ and $\sigma_1 \subset \sigma_e$, $\sigma'_1 \subset \sigma'_e$ then $\sigma_e \prec \sigma'_e$.
Hence

(7.4) If $\sigma_1 \subset \sigma_e$ then $\mathrm{St}\ \sigma_1 \subset \mathrm{St}\ \sigma_e$.

It follows that K'_e is likewise star finite and hence it is locally finite. From (7.4) and the fact that all but a finite number of simplexes of K'_e are in $|K - \mathrm{St}\ \sigma_e|$ it is but a step to proving (7.2) for K'_e.

(7.5) When K_e is finite $|K_e|$ is a

compactum. (AT,VIII,3.1a).

(7.6) When K_e is finite mesh $K_e^{(n)} \longrightarrow 0$.(3.5)

(7.7) THEOREM. Let K be countable and locally finite and let K_e, K_g be an Euclidean and a geometric realization of K. Then the natural barycentric transformation $t : |K_e| \longrightarrow |K_g|$ (see 4.7) is topological.

In short <u>Euclidean complexes are special geometric complexes</u>.

Noteworthy and immediate consequence:

(7.8) All the Euclidean realizations of the same countable locally finite complex K give rise to topologically equivalent polyhedra. More precisely the associated barycentric transformations are topological.

Another proof of (7.8) independent of geometric realizations is found in (AT,VIII,6.4)

(7.9) <u>Proof of (7.7)</u>. The notations remaining those of (4.10,5) identify the points of $|K_g|, |K_e|$ with the same barycentric coordinates and let the corresponding topologies be compared. If $\sigma_n \in K_g^{(n)}$ (barycentric derived throughout), St σ_n is the intersection of the stars of its vertices. Hence $\{\text{St } \sigma_n\}$ (all n, σ_n) is a base for K_g. Let now $\delta(x,x')$ be a distance-function for $|K_e|$ with $\mathfrak{S}(x,\epsilon)$ as the corresponding spheres. Let x_0 be any point of $|K_g|$. Since $L = \text{Cl St } \bar{\sigma}(x_0)$ is finite, by (7.5) mesh $L^{(n)} \longrightarrow 0$. Since St $\sigma_n(x_0)$ is a subcomplex of $L^{(n)}$, given ϵ there is an n such that St $\sigma_n(x_0) \subset \mathfrak{S}(x_0,\epsilon)$.

On the other hand $K_e^{(n)}$ - St $\sigma_n(x_0) = L \cup M$, where $L = K_e^{(n)}$ - St Cl $\sigma_n(x_0)$ and M is a finite set of closed simplexes not containing x_0. Thus $\delta(x_0, |M|) > 0$, and by (7.2) for $K_e^{(n)}$: $\delta(x_0, |L|) > 0$. Hence

$\delta(x_o, |K_e^{(n)} - \text{St}\ \sigma_n(x_o)|) = \epsilon > 0$, and so $\mathfrak{S}(x_o, \epsilon) \subset |\text{St}\ \sigma_n(x_o)|$. Therefore the bases $\{\mathfrak{S}(x, \epsilon)\}$, $\{|\text{St}\ \sigma_n(x)|\}$ are equivalent, which proves the theorem.

(7.10) Every countable locally finite simplicial complex K has an Euclidean realization in the Hilbert parallelotope P^ω. If $n = \dim K$ is finite the realization may be made in any parallelotope P^r, $r \geq 2n+1$. In particular every finite simplicial complex has a realization in some P^r. (AT,VIII,3.4).

(7.11) If K_e, L_e are Euclidean complexes a barycentric transformation $t: |K_e| \longrightarrow |L_e|$ is continuous, i.e. it is a mapping. (AT,VIII,6.1).

In fact t is affine on each $\bar{\sigma}_e$ of K_e, and hence continuous on $\bar{\sigma}_e$. Since each point $x \in |K_e|$ is on a finite number of σ_e, it is an elementary matter to show that if $x_n \rightarrow x$ then $tx_n \longrightarrow tx$.

An immediate consequence of (3.5) and which includes (7.5) is:

(7.12) If mesh $K_e < \lambda$ and $\dim K_e = n$ is finite, then mesh $K_e^{(r)} < \lambda(1 - \frac{1}{n+1})^r$, where the derived is barycentric. Hence mesh $K_e^{(r)} \longrightarrow 0$.

The following property gives a convenient estimate for the displacement under a barycentric mapping.

(7.13) Let K_e, L_e be Euclidean complexes in P^r or P^ω and let t be a barycentric mapping $|K_e| \longrightarrow |L_e|$. Then $\sup d(x,tx)$, $x \in |K_e|$, is reached at a vertex. Thus if the vertices merely undergo an ϵ displacement then t itself is an ϵ displacement.

We may evidently suppose that K_e, L_e are closed Euclidean simplexes $\bar{\sigma}_e$, $\bar{\sigma}'_e$. Suppose first that $p = \dim \sigma_e = 1$, so that σ_e is a segment $l : 0 \leq u \leq 1$. Then $d^2(x,tx) = f(u) = au^2 + bu + c \to +\infty$ when $u \to \pm\infty$. Therefore f has a single extremum which is a minimum on $-\infty < u < +\infty$, and so it reaches its maximum on l at one of the end points. Thus the property holds for $p = 1$ and also trivially for $p = 0$. Grant it for $p - 1$ and let $\sigma_e = A\sigma_e^{p-1}$. If $x \in \bar{\sigma}_e$ there is a segment AB through x in $\bar{\sigma}_e$ with $B \in \bar{\sigma}_e^{p-1}$. By the preceding result $d(x,tx) \leq \sup \{d(A, tA), d(B, tB)\}$. By the hypothesis of the induction $d(B, tB) \leq d(A', tA')$, where A' is a vertex of σ_e^{p-1}. Therefore $d(x, tx) \leq \sup \{d(A, tA), d(A', tA')\}$, proving (7.13).

(7.14) If K_e is finite and possesses a closed subcomplex L_e of mesh $< \epsilon$, then K_e has a simplicial subdivision K_{1e} of mesh likewise $< \epsilon$ and still containing L_e as a subcomplex.

Let us apply n successive derivations modified at each stage in that no new vertices are placed in L_e, which thus remains intact. The only one-simplexes of the finite complex K_{1e} which are not simplexes of $K_e^{(n)}$ are those of the form AB where A is a vertex of a certain $\sigma_e \subset |L_e|$ and B is a point whose distance from the centroid of σ_e tends to zero with increasing n. Since diam AB may be made to exceed diam σ_e as little as possible, n may be so chosen that in K_{1e} all edges are of length $< \epsilon$. Thus (7.14) is a consequence of (3.4).

§3. COMPARISON OF THE TOPOLOGIES ASSOCIATED WITH AFFINE COMPLEXES

8. Diverse topologies may be assigned to the affine representation K_v of a given simplicial complex K. We have first of all the topology induced by the natural metric $d(x,y)$ of K_v (4.11). Since $K_v^{(n)}$ coincides with K_v as a point-set its natural metric $d_n(x,y)(d_o=d)$ is also a metric for K_v and likewise induces its topology. Finally we have the topology of the geometric representation K_g. Following J. Tukey we shall examine the mutual relations existing between these metrics. We shall consider K_v as a point-set to which those various topologies are assigned, so that K_g for example is K_v topologized in a certain special way.

We first require additional properties of the affine simplex $\sigma_v^p = a_o \dots a_p$. A point x of σ_v^p may be represented by its barycentric coordinates in σ_v^p or by those in the simplex of $(\mathrm{Cl}\,\sigma_v^p)'$ containing x. To compare them a more convenient notation for the vertices of the derived is desirable. If $\{j\}$ are the finite subsets of $\{i\} = \{0, 1, \dots, p\}$ then each vertex of $(\mathrm{Cl}\,\sigma_v^p)'$ is associated in one and only one way with a definite j, and this vertex will be designated by a_j. It is important to remember that j <u>is not a special index</u> i. If we set $|j|$ = number of i's in j, the barycentric coordinate of x relative to a_j is denoted by x^j. We now have:

$$(8.1) \qquad x^i = \sum_{j \ni i} |j|^{-1} x^j.$$

(8.2) If $x^{i_1} \geq x^{i_2} \geq \dots \geq x^{i_p}$ then (putting $x^{i_{p+1}} = 0$):

$$x^j = \begin{cases} |j|\,(x^{i_m} - x^{i_{m+1}}) \text{ if } j = \{i_1, \dots, i_m\}, \\ 0 \quad \text{otherwise.} \end{cases}$$

The proof of (8.1) is by an elementary computation which is omitted. We then verify (8.2) by substitution in (8.1).

(8.3) In the natural metric mesh $(\mathrm{Cl}\ \sigma^p_v)^{(n)} \longrightarrow 1$ for fixed n and increasing p.

A one-simplex of $(\mathrm{Cl}\ \sigma^p_v)'$ is of the form $\sigma_1 = \hat{\sigma}^q\hat{\sigma}^r$, $\sigma^q \prec \sigma^r$. Hence if $x \in \sigma_1$ then $x^i = \frac{1}{q+1}$ for $a_i < \sigma^q$, $x^i = \frac{1}{r+1}$ for $a_i \prec \sigma^r$, and $x^i = 0$ otherwise. Therefore if $x, y \in \sigma_1$ then

$$d(x,y) = \left[\frac{r-q}{(r+1)^2} + (q+1)\left(\frac{1}{q+1} - \frac{1}{r+1}\right)^2\right]^{1/2}$$

$$= \left[\frac{r-q}{(q+1)(r+1)}\right]^{1/2} < 1.$$

Therefore diam $\sigma_1 < 1$ and so by (3.4) mesh $(\mathrm{Cl}\ \sigma^p_v)^{(n)} \leqq 1$.

Let now a_{1i} denote the centroid of $a_o \ldots a_i$, and having defined $a_{ro}, a_{r1}, \ldots$ let $a_{r+1,i}$ denote the centroid of $a_{ro} \ldots a_{ri}$. Then $\sigma^1_n = a_o\, a_{np}$ is a one-simplex of $(\mathrm{Cl}\ \sigma^p_v)^{(n)}$ and if the barycentric coordinates of x_{np} in σ_p are x^i_{np} we have:

$$\operatorname{diam}\ \sigma^1_n = \left[(x^o_{np} - 1)^2 + \sum_{i > 0} (x^i_{np})^2\right]^{1/2} > 1 - x^o_{np}.$$

Therefore by (3.4):

$$1 - x^o_{np} \leqq \text{mesh } (\mathrm{Cl}\ \sigma^p_v)^{(n)} \leqq 1.$$

Now $x^o_{1p} = \frac{1}{p+1}$ and in general the x^o_{np} for fixed n are the Cesaró means of order n of the sequence $1, 0, 0, \ldots$ which $\longrightarrow 0$. Since Cesaró means preserve convergence under the same conditions $x^o_{np} \longrightarrow 0$ and so (8.3) follows.

(8.4) If every $|x^{i\prime} - x^{i\prime\prime}|$ is either 0 or $> \epsilon$ and if $\sum |x^i - y^i| < \epsilon$ then for any j

$$|x^j - y^j| \leqq |j| \sum |x^i - y^i|.$$

Let $y^{i_1} \geq y^{i_2} \geq \dots \geq y^{i_n}$ include all the non-zero coordinates of y. Then for $i_o \neq i_k$ and any $k : |x^{i_o}| \leq |x^{i_o} - y^{i_o}| < \epsilon$ and hence under the assumptions of (8.4) $x^{i_o} = 0$. Then also

$$x^{i_{k+1}} - x^{i_k} \leq y^{i_{k+1}} - y^{i_k} + |y^{i_{k+1}} - x^{i_{k+1}}| + |y^{i_k} - x^{i_k}|$$

$$\leq 0 + \sum |y_i - x_i| \leq \epsilon.$$

Hence $x^{i_{k+1}} \leq x^{i_k}$, or $x^{j_1} \geq x^{i_2} \geq \dots \geq x^{i_n}$ and this includes all non-zero x^i. Hence by (8.2):

$$|x^j - y^j| \leq |j| \, \{|x^{i_j} - y^{i_m}| + |x^{i_{m+1}} - y^{i_{m+1}}|\}$$

$$\leq |j| \sum |x^i - y^i|$$

in the only cases $j = (i_1, \dots , i_m)$ where we do not have $x^j = y^j = 0$ in (8.2).

9. The preceding results will now be applied to our representations. Referring $|K_v|$ to its natural metric $d(x,y)$ we define as usual mesh $K_v^{(p)} = \sup \operatorname{diam} \sigma^p$ in this metric and likewise for its subcomplexes. And now from (3.4) and (8.3) there comes:

(9.1) If dim $K = n$ (finite) then mesh $K_v^{(p)} \leq (\frac{n}{n+1})^p$ mesh K_v which $\to 0$.

(9.2) If dim $K = \infty$ then mesh $K_v^{(p)} = 1$ for all p.

(9.3) Mesh St $\sigma_p(x) \to 0$ for every x, if and only if the dimension of the star of each vertex of K is finite.

We pass to the comparison of the distances $d_n(x,y)$. We shall denote by σ not only a given simplex but also the set of indices of its vertices. Thus $i \in \sigma$ signifies that a_i is a vertex of σ. We designate temporarily by $|\sigma|$ the number $1 + \dim \sigma$.

(9.4) If $x \in \sigma$ and y is any other point then:

$$|x^i - y^i| \leq d(x,y) \leq \sum |x^i - y^i| \leq 2\sum_{i \in \sigma} |x^i - y^i|$$

$$\leq 2|\sigma| d(x,y).$$

The only inequality which is not obvious is the one between the third and fourth terms. Now

$$\sum_{i \notin \sigma} |x^i - y^i| = \sum_{i \notin \sigma} y_i = 1 - \sum_{i \in \sigma} y_i$$

$$= \sum_{i \in \sigma} x_i - \sum_{i \in \sigma} y_i \leq \sum_{i \in \sigma} |x^i - y^i|$$

which implies the desired result.

Let l_i denote the segment $0 \leq x_i \leq 1$ and let $Q = \prod l_i$. By means of the transformation sending $x \in |K_v|$ into the point of Q whose coordinates are the barycentric coordinates of x, we may identify K_v with a subset of Q and $|K_v|$ acquires thus a certain relative topology. From (9.4) there comes

(9.5) The relative topology of $|K_v|$ when immersed in Q agrees with the topology induced by $d(x,y)$.

Suppose $x \in \sigma$ and ϵ given and let $U = \{y \mid |x^i - y^i| < \frac{\epsilon}{4|\sigma|}, i \in \sigma\}$. By (9.4) we then have $d(x,y) < \epsilon$ and so $U \subset \mathfrak{S}(x, \epsilon)$. Thus there is an open set of the relative topology between x and a given $\mathfrak{S}(x, \epsilon)$. Conversely given any open set V of the relative topology containing x there is between x and V an open set U consisting of the points y such that $|x^i - y^i| < \epsilon$ for $i = i_1, \ldots, i_r$. By (9.4) if $d(x,y) < \epsilon$ then $y \in U$ and hence $y \in V$, or $\mathfrak{S}(x, \epsilon) \subset V$, and (9.5) follows.

Let t be a mapping of a topological space $\mathcal{R} \to |K_v|$. We have then $x = t\xi = (t^i\xi)a_i$, where $t^i\xi$ denotes the x^i coordinate of $t\xi$. Then from (9.5) and (AT,I,12.1, 12.2) there follows:

(9.6) t is continuous if and only if each t^i is continuous.

By means of (8.4) we derive:

(9.7) For each $x \in |K_v|$ there is a finite N_x such that

$$\frac{1}{N_x} d(x,y) \leqq d_1(x,y) \leqq N_x \, d(x,y).$$

Since x is fixed and only a finite number of $x^i \neq 0$ we may choose ϵ to satisfy the hypothesis of (8.4). Let a_i be the vertices of $\sigma(x)$ and a_j those of $\sigma_1(x)$. Then we may take

$$N_x = \sup \left\{ \frac{4|\sigma|}{\epsilon}, \; 4|\sigma| \sum_{j \in \sigma_1} |j| \right\}.$$

For if $d(x,y) \geqq \frac{\epsilon}{2|\sigma|}$ then $N_x \, d(x,y) \geqq 2$. Since

$$d_1(x,y) \leqq \left[(\sum x^j)^2 + (\sum y^j)^2 \right]^{1/2} \leqq \sqrt{2},$$

the right hand inequality of (9.7) will hold. And on the other hand if $d(x,y) < \frac{\epsilon}{2|\sigma|}$ then by (9.4): $\sum |x^i - y^i| \leqq 2|\sigma| d(x,y) < \epsilon$, so we may apply (8.4) to obtain:

$$d(x,y) \leqq \sum |x^j - y^j| \leqq 2 \sum_{j \in \sigma_1} |x^j - y^j|$$

$$\leqq 2 \sum_{j \in \sigma_1} |j| \sum_i |x^i - y^i|$$

$$\leqq 2 \Big(\sum_{j \in \sigma_1} |j| \Big) \cdot 2|\sigma| d(x,y) \leqq N_x \, d(x,y).$$

Therefore the right-hand inequality of (9.7) holds again and so it holds in all cases.

Now we have by (8.1):

$$d(x,y) \leqq \sum |x^i - y^i| \leqq \sum_i \sum_{j \ni i} |x^j - y^j|$$
$$= \sum |x^j - y^j| \leqq 2|\sigma_1| d_1(x,y) \leqq N_x\, d_1(x,y),$$

since $2|\sigma_1| \leqq 2 \sum_{j\in\sigma_1} |j| \leqq N_x$. Therefore the left-hand inequality of (9.7) holds also.

It is evident from (9.7) that if one of $d(x,y)$, $d_1(x,y) \to 0$ so does the other. Hence $d(x,y)$ and $d_1(x,y)$ induce the same topology in $|K_v|$. Consequently:

(9.8) All the metrics $d_n(x,y)$ induce the same topology in $|K_v|$.

If a_i is a vertex of K_v then the points for which $x^i > 0$ are precisely those of St a_i and since by (9.6) the barycentric coordinates are continuous in $|K_v|$, $|\text{St } a_i|$ is an open set of $|K_v|$. Since the same argument applies to the stars of the vertices of $K_v^{(n)}$ we have:

(9.9) The open sets of $|K_g|$ are likewise open sets of $|K_v|$.

From (9.3) follows on the other hand that the open sets of $|K_v|$ are open in $|K_g|$ when and only when all the St a_i are finite dimensional. Therefore we have the comprehensive

(9.10) THEOREM. The topologies induced by the natural metrics of K_v, K_v', ... all agree with one another, and they agree likewise with the topology of K_g when and only when all the stars of the vertices of K are finite dimensional.

In particular, and this is the case which interests us especially in the sequel:

(9.11) COROLLARY. When K is locally finite the topologies induced by the metrics of K_v, K_v' ... , and the topology of K_g are all in agreement. Thus when K_e is an Euclidean complex its metric induces the same topology as the natural metrics of the affine complexes K_v, K_v', ...

Chapter II

SINGULAR COMPLEXES

1. An extensive discussion of singular complexes has already been given in (AT,VIII,24-27) with particular emphasis upon the applications to polyhedra. We return to the question chiefly to add certain complements required in connection with the topics taken up later in (IV).

2. (2.1) We recall (AT,VIII,24) that a singular p-cell in the metric space $\mathfrak{R}$ is a pair $E^p = (\sigma^p, s)$, where σ^p is an Euclidean simplex and s a mapping $\bar{\sigma}^p \longrightarrow \mathfrak{R}$. It is understood that if s' is a barycentric mapping $\bar{\sigma}^p \longrightarrow \bar{\sigma}'^p$ then also $E^p = (s'\sigma^p, ss'^{-1})$. For details regarding the complete singular complex $\Sigma = \{E\}$ of $\mathfrak{R}$ see (AT,VIII,24).

Let now $\mathfrak{R}$ be a compactum and Ω its complete Vietoris complex (AT,VII,23). Since Σ is a metric complex we may introduce its V-cycles (analogues of the Vietoris cycles; see AT,VI,25) and they will be referred to as the VS-<u>cycles</u> of $\mathfrak{R}$. Although they do not yield new groups, they will nevertheless be very convenient auxiliaries later.

(2.2) With E^p as before let $\sigma^p = A_0 \ldots A_p$. The sA_i are the vertices of a simplex of Ω which is written $\sigma(E)$. If $\dim \sigma(E) = p$, i.e. if the vertices of E are all distinct, we choose explicitly $\sigma(E) = (sA_0) \ldots (sA_p)$. Now $E \rightarrow \sigma(E)$ defines a closed set-transformation $t : \Sigma \longrightarrow \Omega$ (AT,IV,7). We associate with t the chain-transformation $\tau : \Sigma \to \Omega$ with the closed carrier t defined by: $\tau E = \sigma(E)$ whenever $\dim \sigma(E) = \dim E$, $\tau E = 0$ otherwise. That τ is a chain-mapping is proved thus: Suppose E^p as above and let s induce the chain-mapping $\theta : \sigma^p \to \Sigma$. Then we have to prove: $\tau F \theta\ \sigma^p = F\tau\theta\ \sigma^p$. Since θ is a chain-mapping this reduces to $(\tau\theta)F = F(\tau\theta)$, which is obvious since $\tau\theta$ is a simplicial chain-mapping $\mathrm{Cl}\,\sigma \to \Omega$. It is clear also that τ is an f chain-mapping (AT,IV, 19). Finally diam $tE \leq$ diam E, and so t is metric.

If $\mathfrak{R}$ is a singular complex and C a singular chain we shall say that $\mathfrak{R}$ is <u>adherent</u> to $t\mathfrak{R}$ and C <u>adherent</u> to τC. We may also describe t as <u>set-adherence</u> and τ as <u>chain-adherence</u>, and the pair (t, τ) as an <u>adherence</u>.

Coupling the results just proved with (AT,VI,27.2) we have:

(2.3) Set-adherence t is a metric set-transformation $\Sigma \to \Omega$ and chain-adherence τ a metric chain-mapping $\Sigma \to \Omega$. Hence τ induces a homomorphism of the homology groups of the VS-cycles of $\mathfrak{R}$ into the corresponding Vietoris groups. Classes which correspond under this homomorphism are said to be adherent.

3. Let $\mathfrak{R}$, $\mathfrak{R}_1$ be metric spaces and Σ, Σ_1 their complete singular complexes. Suppose that there is a mapping $T: \mathfrak{R} \to \mathfrak{R}_1$. If $E^p = (\sigma^p, s)$ is a singular p-cell of $\mathfrak{R}$ then $E_1^p = (\sigma^p, Ts)$, denoted by TE^p, is a singular p-cell of $\mathfrak{R}_1$. Since $E' \prec E \Rightarrow TE' \prec TE$, T defines a set-transformation $t: \Sigma \to \Sigma_1$ (AT,IV,7.1). Moreover $E \to TE$ defines an f chain-transformation $\tau: \Sigma \to \Sigma_1$ which is a chain-mapping. For $\tau F = F\tau$ reduces to $\tau FE^p = F\tau E^p$, and this is a consequence of

the definition of F for Σ. It is clear that τ has the closed carrier t, which is single-valued. We notice also that if $\mathfrak{R}$, $\mathfrak{R}_1$ are compacta then both t and τ are metric.

Referring to (AT,IV,9.5,19.3) and (AT,VI,27.2) we have then:

> (3.1) A mapping T: $\mathfrak{R} \longrightarrow \mathfrak{R}_1$ induces a closed set-transformation t: $\Sigma \longrightarrow \Sigma_1$, as well as a chain-mapping τ: $\Sigma \longrightarrow \Sigma_1$ with t for carrier, both being metric when $\mathfrak{R}$, $\mathfrak{R}_1$ are compacta. Hence T induces also homomorphisms of the homology groups: (a) of the finite singular cycles of $\mathfrak{R}$ into the corresponding groups of $\mathfrak{R}_1$; (b) likewise of those of the VS-cycles of $\mathfrak{R}$ into the corresponding groups of $\mathfrak{R}_1$ if both $\mathfrak{R}$ and $\mathfrak{R}_1$ are compacta.

If T is topological t is an isomorphism and the homomorphisms in (3.1) become isomorphisms also. Moreover if $\mathfrak{R}$, $\mathfrak{R}_1$ are compacta t is a metric isomorphism. Therefore:

> (3.2) The homology groups of the finite singular cycles of a metric space $\mathfrak{R}$, and when $\mathfrak{R}$ is a compactum, those of its VS-cycles, are topologically invariant.

4. (4.1) Let $\mathfrak{K} = (K,t)$, $K = \{\sigma\}$, be a continuous complex in $\mathfrak{R}$ and introduce the singular cells $E = (\sigma, t)$. By (3.1) (with K considered as a singular complex in the polyhedron $|K|$) we find that $\tilde{\mathfrak{K}} = \{E\}$ is a singular complex in $\mathfrak{R}$ with an attached chain-mapping $\tau : K \longrightarrow \tilde{\mathfrak{K}}$ such that $\tau\sigma = E$. We call $\tilde{\mathfrak{K}}$ the <u>singular complex</u> induced by $\mathfrak{K}$. It is merely the singular complex composed of the continuous (= singular) cells

of $\mathfrak{K}$.

(4.2) Conversely as shown in (AT,VIII,26.4) every singular complex $\tilde{\mathfrak{K}}$ is induced by at least one continuous complete $\mathfrak{K}$, referred to as an antecedent of $\tilde{\mathfrak{K}}$. Notice that if $\tilde{\mathfrak{K}}$ is locally finite (and this is the only case of interest) then the construction loc. cit. yields an antecedent $\mathfrak{K}$ in which each cell of $\tilde{\mathfrak{K}}$ has only a finite number of antecedents.

(4.3) Let then $\tilde{\mathfrak{K}}$ be locally finite and $\mathfrak{K}$ = (K,t) an antecedent such that for every E_i^p the set $\{\sigma_{ih}^p\}$ of simplexes of K, with the property that $E_i^p = (\sigma_{ih}^p, t)$, is finite. Then if $C^p = x^{ih}\,\sigma_{ih}^p$ is a chain of K there corresponds to it the chain $D^p = y^i\, E_i^p$, $y^i = \sum x^{ih}$, which we may denote by $D^p = (C^p,t)$. We refer conveniently to D^p as a (singular) chain of $\mathfrak{K}$. This relationship, however, carries but little weight. For let D^p be any singular chain such that the singular complex $|D^p|$ is locally finite, say the chain $D^p = y^i\, E_i^p$ of $\tilde{\mathfrak{K}}$. This is no restriction on D^p since $\tilde{\mathfrak{K}}$ is merely any locally finite singular complex and could as well be $\|D^p\|$ itself. Then $D^p = (y^i\,\sigma_{i1}^p, t)$, and so D^p is a chain of $\mathfrak{K}$ in the above sense. Thus for instance every finite singular chain is a chain of some continuous complex.

The situation is entirely different regarding the cycles. Since $\sigma_{ih}^p \to E_i^p$ defines a chain-mapping τ: $K \to \tilde{\mathfrak{K}}$, if γ^p is a cycle of K then $\delta^p = (\gamma^p, t)$ is a cycle of $\tilde{\mathfrak{K}}$. However the converse need not be true (as simple examples show) and there may exist cycles δ^p of $\tilde{\mathfrak{K}}$ which are not of the form (γ^p, t). For this reason we lay down the

(4.4) DEFINITION. A singular cycle δ^p is said to be a continuous cycle of $\mathfrak{R}$ whenever there exists a continuous complex (K,t) and a cycle γ^p of K such that $\delta^p = (\gamma^p,t)$

It is important to bear in mind that <u>a continuous cycle δ^p is also a singular cycle</u>. This means for instance that the antecedent (K,t) may be considerably modified without affecting δ^p. Thus suppose that $|K_1| = s\,|K|$, where s is simplicial barycentric. Then we still have $\delta^p = (s\,\gamma^p, ts^{-1})$. As a consequence we may state:

(4.5) The finite continuous cycles over a discrete G form a subgroup $\mathfrak{Z}_c^p(\mathfrak{R},G)$ of the group of the singular cycles $\mathfrak{Z}_s^p(\mathfrak{R},G)$. Similarly for the related $\mathfrak{F}$ groups of the bounding cycles, and hence we may introduce the homology groups: $\mathfrak{H}_c^p(\mathfrak{R},G) \subset \mathfrak{H}_s^p(\mathfrak{R},G)$.

As it happens in all the interesting applications $\mathfrak{H}_c^p \cong \mathfrak{H}_s^p$, and so there will be no occasion to consider $\mathfrak{H}_c^p$ separately.

The proof of (4.5) is elementary. Suppose $\delta_1^p = (\gamma_1^p, t_1)$, $\delta_2^p = (\gamma_2^p, t_2)$ continuous. We may choose γ_1^p, γ_2^p in disjoint Euclidean complexes K_1, K_2 in some parallelotope. Then if $K = K_1 \cup K_2$ and $t : |K| \to \mathfrak{R}$ is defined by $t|\,|K_1| = t_1$, we have $\delta_1^p = (\gamma_1^p, t), \gamma_1^p$ in K, and so $\delta_1^p - \delta_2^p = (\gamma_1^p - \gamma_2^p, t)$, proving the asserted property for the groups $\mathfrak{Z}_c^p$. It is also clear that if $\delta^p = (\gamma^p, t)$ then $F\delta^p = (F\gamma^p, t)$ and so $F\,\mathfrak{Z}_c^p \subset \mathfrak{Z}_c^{p-1}$, from which the rest of (4.5) follows.

(4.6) A continuous complex $\mathfrak{K} = (K,t)$, $K = \{\sigma\}$, is said to be <u>simplicial</u> whenever: (a) (σ^p, t) always has $p + 1$ distinct vertices; (b) if $\sigma \neq \sigma'$ then (σ,t) and (σ',t) do not have the same set of vertices. Similarly a singular complex $\mathfrak{K} = \{E\}$ is said to be <u>simplicial</u> whenever: (a) E^p always has $p + 1$ vertices; (b) if $E \neq E'$ then they do not have the same set of vertices. A singular chain C^p is said to be <u>simplicial</u> if the singular complex $|C^p|$ is simplicial.

(4.7) A finite singular simplicial complex $\mathfrak{K} = \{E\}$ has an antecedent (K, t) such that if $K = \{\sigma\}$ and $E = (\sigma, t)$, then $\sigma \rightarrow E$ defines an isomorphism $K \rightarrow \mathfrak{K}$. Hence: (a) the cycles of $\mathfrak{K}$ are all continuous; (b) finite simplicial cycles are continuous cycles.

Let $q = \dim \mathfrak{K}$. The assertion is trivial for $q = 0$, so we assume it for $q - 1$ $(q > 0)$ and prove it for q. By hypothesis the $(q-1)$-section $\mathfrak{K}^{q-1}$ of $\mathfrak{K}$ has for antecedent a continuous $(q-1)$-complex (K^{q-1}, t). Let $\{A_i\}$, $\{B_i\}$ be the vertices of $\mathfrak{K}$, K^{q-1}, and let them be so labelled that $A_i = tB_i$. Since $|K^{q-1}|$ may be replaced by any topological barycentric isomorph, we may suppose $|K^{q-1}| \subset P^{\omega}$ and such that its vertices are independent. In particular any $q + 1$ will be the vertices of a σ^q. Consider now any E^q and let the notations be so chosen that E^q has the vertices $A_0, \ldots,$ A_q. Then if $\sigma^q = B_0 \ldots B_q$, we have $E_q = (\sigma^q, t')$, where t' is defined on $|\mathfrak{B}\sigma^q|$ and in fact $t' \,|\, |\mathfrak{B}\dot{\sigma}^q| = t \,|\, |\mathfrak{B}\sigma^q|$. It follows that we may extend t to σ^q by choosing $t = t'$ on σ^q. If the σ^q are added to K^{q-1}. there is obtained a complex K, related to $\mathfrak{K}$ as in (4.7).

(4.8) An important consequence of (4.7) is that in dealing with a singular simplicial complex $\mathfrak{K}$ it will not be necessary to distinguish between $\mathfrak{K}$ and the associated (K, t). We shall thus speak indifferently of the chains, cycles, groups of (K, t), meaning thereby those of $\mathfrak{K}$.

Remark. One would be tempted to replace general singular complexes and chains by the simplicial types. However: (a) if $\mathfrak{K}$, $\mathfrak{L}$ are simplicial $\mathfrak{K} \cup \mathfrak{L}$ need not be; (b) the image of a simplicial $\mathfrak{K}$ under a mapping of the space $\mathcal{R}$ need not be simplicial.

Thus simplicial singular complexes have deficiencies which limit their application.

5. Returning to the antecedents, they will enable us to define certain operations in singular complexes. To simplify matters let the singular complex $\tilde{\mathfrak{K}}$ be locally finite and let otherwise the notations remain as in (4.1, 4.2, 4.3). Operations whose effect on the σ_{ih} is independent of h will define operations in $\tilde{\mathfrak{K}}$ itself. It is convenient to adopt a rather general formulation. Let $\tilde{\mathfrak{K}}_1, \ldots$ be analogues of $\tilde{\mathfrak{K}}, \ldots$ and let τ be a chain-mapping $K \to K_1$ with the property that if σ', σ'' are both antecedents of $E \in \mathfrak{K}$, then $C = (\tau\sigma', t_1) = (\tau\sigma'', t_1)$. In other words C is a singular chain of $\tilde{\mathfrak{K}}_1$ which depends solely upon E. Then

> (5.1) The chain-transformation $\theta : \tilde{\mathfrak{K}} \to \tilde{\mathfrak{K}}_1$ defined by $\theta\ E = C$ is a chain-mapping. It is said to be induced by τ.

We have to show that $\theta\, FE = F\, \theta\, E$. Now $F\, \theta\, E = F(\tau\, \sigma, t_1) = (\tau F\sigma, t_1)$, since τ is a chain-mapping. On the other hand the definition of θ yields $\theta\, FE = (\tau F\sigma, t_1) = F\, \theta\, E$, proving (5.1).

Applications (5.2). Take $K_1 = K^{(n)}$(nth barycentric derived), and $t_1 = t$, so that $\mathfrak{K}_1 = \mathfrak{K}^{(n)}$. The singular complex $\tilde{\mathfrak{K}}_1$ induced by $\mathfrak{K}_1$ is written $\tilde{\mathfrak{K}}^{(n)}$ and called the nth derived of $\tilde{\mathfrak{K}}$. Let δ denote chain-derivation in K and let σ, σ' be antecedents of the same E. There exists then a topological barycentric mapping $s: \bar{\sigma}' \to \bar{\sigma}''$ such that if $t|\bar{\sigma}' = t'$, $t|\bar{\sigma}'' = t''$ then $t'' = st'$. We find at once from the definition of δ given in (AT,IV,27), that $\delta\sigma'' = s\delta\sigma'$, and this yields $(\delta\sigma', t') = (\delta\sigma'', t'')$, or $(\delta\sigma', t) = (\delta\sigma'', t)$. Therefore δ induces a singular chain-mapping $d: \tilde{\mathfrak{K}} \to \tilde{\mathfrak{K}}'$ known as singular chain-derivation in $\tilde{\mathfrak{K}}$. By repetition we obtain. $d^n: \tilde{\mathfrak{K}} \to \tilde{\mathfrak{K}}^{(n)}$ induced by δ^n.

It will be observed that if $\mathfrak{K}$ is finite then mesh $K^{(n)} \to 0$ and therefore also mesh $\tilde{\mathfrak{K}}^{(n)} \to 0$. Thus

n may be so chosen that mesh $\tilde{\mathfrak{K}}^{(n)} < \epsilon$ assigned.

(5.3) Let γ^p be a finite singular cycle of mesh $< \epsilon$, and suppose that $\gamma^p = FC^{p+1}$ where C^{p+1} is a finite singular chain. Choosing this time $\tilde{\mathfrak{K}} = \|C^{p+1}\|$, there is a closed subcomplex L of K such that (L,t) induces $\|\gamma^p\|$. By (I, 7.15) if mesh $L < \eta$ there is a subdivision K_1 of K whose mesh $< \eta$ and such that L is still a subcomplex of K_1. If δ' is the related chain-subdivision and d' the operation induced in $\tilde{\mathfrak{K}}$, then for suitable K_1 we will have mesh $\tilde{\mathfrak{K}}_1 < \epsilon$, and $d'\gamma^p = \gamma^p$. Since d' is a chain-mapping we have $d'\,FC^{p+1} = F(d'\,C^{p+1}) = \gamma^p$. Thus γ^p bounds the chain $d'\,C^{p+1}$ of $\tilde{\mathfrak{K}}_1$ and hence in the set $\|C^{p+1}\|$ whose mesh $< \epsilon$, and so:

> (5.4) If a singular cycle γ^p of mesh $< \epsilon$ bounds a singular chain C^{p+1} then it bounds in the set $\|C^{p+1}\|$ a singular chain whose mesh $< \epsilon$.

6. In connection with homotopy it is convenient to enlarge the set of singular cells by introducing <u>singular prismatic cells</u>. We understand thereby a pair $e = (l \times \sigma, t)$, $l : 0 \leq u \leq 1$, t a mapping $l \times \bar{\sigma} \to \mathcal{R}$, and it is agreed that $e = (l \times \sigma', t')$ if $\sigma' = s^{-1}\sigma$, s barycentric topological, and $t'(u \times x') = t(u \times s\,x')$, $x' \in \sigma'$.

If $E = (0 \times \sigma, t)$, $E' = (1 \times \sigma, t')$ we will denote e by $\mathfrak{D}E$ and set $\mathfrak{D}\Sigma = \{E, \mathfrak{D}E\}$ = the collection of all the singular cells and singular prismatic cells. We turn $\mathfrak{D}\Sigma$ into a complex in the obvious way. The dimensions and incidence relations in Σ are as before. Then we define $\dim \mathfrak{D}E = 1 + \dim E$, $\mathrm{Cl}\ \mathfrak{D}E = \mathrm{Cl}\ E \cup \mathrm{Cl}\ E' \cup \{(l \times \sigma', t) \mid \sigma' \prec \sigma\}$. Finally the incidence-numbers are chosen recursively as the numbers making the basic relation

(6.1) $$F\mathfrak{D}E = E' - E - \mathfrak{D}FE$$

hold throughout.

(6.2) Let $\overline{1\times\sigma}$ undergo a barycentric subdivision modified in that no new vertices are placed on the elements $0\times\sigma$, $1\times\sigma$, and let Δ be the associated chain-subdivision. If $\Delta(1\times\sigma) = x^1\sigma_1$ we define $\delta\mathfrak{D}E = \mathfrak{D}_1 E = x^1(\sigma_1, t)$. Since Δ is a chain-mapping we have:

$F\Delta(1\times\sigma) = \Delta F(1\times\sigma) = 1\times\sigma - 0\times\sigma - \Delta(1\times F\sigma)$,

and hence

$$F\mathfrak{D}_1 E = E' - E - \mathfrak{D}_1 FE.$$

A chain-homotopy of $\mathfrak{K}$, $\mathfrak{K}_1$ in $\mathfrak{D}\Sigma$ with deformation-chains $\mathfrak{D}E$ is said to be <u>singular-prismatic</u>. The corresponding chain-homotopy in Σ itself with deformation-chains $\mathfrak{D}E$ is said to be <u>singular</u>. We have just proved in substance:

(6.3) A singular prismatic chain-deformation $\tau: \mathfrak{K}\to\mathfrak{K}_1$ is likewise a singular chain-deformation, and if $\mathfrak{D}$, $\mathfrak{D}_1$ are the corresponding operators then $\|\mathfrak{D}E\| = \|\mathfrak{D}_1 E\|$ for every $E\in\mathfrak{K}$.

(6.4) Let $\mathfrak{R}$, $\mathfrak{R}'$ be metric spaces and Σ, Σ' their complete singular complexes. If T_1, T_2 are homotopic mappings $\mathfrak{R}\longrightarrow\mathfrak{R}'$ then the corresponding induced chain-mappings τ_1, τ_2 are singularly chain-homotopic.

(6.5) Let $\mathfrak{D}'\Sigma'$, $\mathfrak{D}_1'$ be the analogues of $\mathfrak{D}\Sigma$, $\mathfrak{D}_1$ for $\mathfrak{R}'$. It is an elementary consequence of the homotopy, proved like (3.1), that there is a chain-mapping $\theta: 1\times\Sigma\longrightarrow\mathfrak{D}'\Sigma'$ such that $\theta(0\times E) = \tau_1 E = E'$, $\theta(1\times E) = \tau_2 E = E''$; $\theta(1\times E) = \mathfrak{D}'E'$, where $E', E''\in\Sigma'$ and $\mathfrak{D}'E'\in\mathfrak{D}'\Sigma'$. Since θ is a chain-mapping the preservation of the boundary relations in $1\times\Sigma$ yields

$$F\,\mathfrak{D}'E' = E'' - E' - \mathfrak{D}'FE'.$$

Passing as before to $\mathfrak{D}_1'$, and writing $\Delta E = \mathfrak{D}_1'E'$, we find

$$F\Delta E = (\tau_2 - \tau_1 - \Delta F)E.$$

Since ΔE is a chain of Σ' this proves (6.4).

As an obvious application we have:

(6.6) Let $(1 \times K, T)$ be a continuous prism in $\mathcal{R}$ such that $(0 \times K, T)$ induces $\mathfrak{K}$ and $(1 \times K, T)$ induces $\mathfrak{K}_1$. Suppose that whenever $\sigma', \sigma'' \in K$ are such that $(0 \times \sigma', T), = (0 \times \sigma'', T) = E \in \mathfrak{K}$, then $(1 \times \sigma', T) = (1 \times \sigma'', T) = E'$, and also $(l \times \sigma', T) = (l \times \sigma'', T) = \mathfrak{D}E$. Then $E \longrightarrow E'$ defines a singular chain-deformation $\mathfrak{K} \longrightarrow \mathfrak{K}_1$ in $\mathcal{R}$.

We also prove:

(6.7) d^n is a singular chain-deformation $\mathfrak{K} \longrightarrow \mathfrak{K}^{(n)}$ with operator $\mathfrak{D}$ such that $\|\mathfrak{D}E\| \subset \overline{|E|}$, $E \in \mathfrak{K}$.

Since $(6.7)_n$ follows from $(6.7)_1$ by repetition, it is sufficient to consider $n = 1$. Let $E = (\sigma, t)$ and let l be as before. Consider a mapping $T : l \times \bar{\sigma} \longrightarrow \mathcal{R}$ such that $T(l \times x) = tx$. Apply to $l \times \bar{\sigma}$ a simplicial partition which differs from a barycentric subdivision only in that no new vertices are inserted in $0 \times \bar{\sigma}$, and let η be the corresponding chain-subdivision. We have then for every face σ_1 of σ:

$$F(l \times \sigma_1) = 0 \times \sigma_1 - 1 \times \sigma_1 - l \times (F\sigma_1).$$

Recalling that η commutes with F and that $\eta | 0 \times \mathrm{Cl}\,\sigma = 1, \eta | 1 \times \mathrm{Cl}\,\sigma = \delta | 1 \times \mathrm{Cl}\,\sigma$, we find

$$F\eta(1 \times \sigma_1) = 1 \times \delta\sigma_1 - 0 \times \sigma_1 - \eta 1 \times F\sigma_1 .$$

Passing now to the singular images, and denoting by $\mathfrak{D}E$ the singular image of $\eta(1 \times \sigma)$, we find that $\mathfrak{D}E$ depends solely upon E and not upon the special representation (σ, t) of the cell. The relation just obtained yields then with $E_1 = (\sigma_1, t)$:

$$F\mathfrak{D}E_1 = d\,E_1 - E_1 - \mathfrak{D}F\,E_1 .$$

Thus d is a singular chain-deformation with operator $\mathfrak{D}$. Since $\|\mathfrak{D}E\| = T(1 \times \sigma) \subset \overline{|E|}$, (6.7) follows.

7. APPLICATION TO HOMOLOGY. (7.1) Every finite singular cycle γ^p is homologous on $\|\gamma^p\|$ to a cycle of arbitrarily small mesh.

Let $\mathfrak{K} = |\gamma^p|$. Given $\epsilon > 0$ there is an n such that mesh $\mathfrak{K}^{(n)} < \epsilon$. If d denotes chain-derivation we have mesh $d^n\gamma^p < \epsilon$. By (6.7) d^n is a singular chain-deformation over $\|\gamma^p\|$ sending γ^p into $d^n\gamma^p$. Hence $\gamma^p \sim d^n\gamma^p$ in $\|\gamma^p\|$, proving (7.1).

(7.2) Every homology class of singular finite cycles contains cycles of arbitrarily small mesh (7.1).

Let $\mathfrak{H}_s$, $\mathfrak{H}_{vs}$ denote the homology groups of the finite singular cycles, and of the VS-cycles. The results just obtained provide us with the means to prove the important

(7.3) THEOREM. $\mathfrak{H}^p_s(\mathfrak{R}, G) \cong \mathfrak{H}^p_{vs}(\mathfrak{R}, G)$.

We shall denote the respective elements of the two groups by Γ^p_s and Γ^p_{vs}. Let then $\{\gamma^p_n\} \in \Gamma^p_{vs}$. It follows at once from the definitions that all the γ^p_n for all the γ^p in the same class are in the same Γ^p_s and that τ: $\Gamma^p_{vs} \rightarrow \Gamma^p_s$ is a homomorphism. Suppose $\tau\Gamma^p_{vs} = \Gamma^p_s = 0$. Then $\gamma^p_n = FC^{p+1}_n$, where by (5.4) we may choose C^{p+1}_n such that mesh $C^{p+1}_n < 2$ mesh $\gamma^p_n + \frac{1}{n} \rightarrow 0$. Therefore $\Gamma^p_{vs} = 0$. Thus τ is univalent. Consider now any Γ^p_s and $\gamma^p \in \Gamma^p_s$. Denoting again chain derivation by d, we have by (6.7): $d^{n+1}\gamma^p \sim d^n\gamma^p$. Hence there is a C^{p+1}_n of mesh $<$ mesh $d^n\gamma^p$ + mesh $d^{n+1}\gamma^p + \frac{1}{n} \rightarrow 0$, such that $FC^{p+1}_n = d^{n+1}\gamma^p - d^n\gamma^p$. Since mesh $d^n\gamma^p \rightarrow 0$ we have in $\{d^n\gamma^p\}$ a VS-cycle whose class Γ^p_{vs} is such that $\tau\Gamma^p_{vs} = \Gamma^p_s$. Therefore τ is a univalent mapping onto, and so it is an isomorphism. This proves (7.3).

Chapter III

MAPPING AND IMBEDDING THEOREMS. RETRACTION

The present chapter deals chiefly with certain general mapping properties. There are two basic theorems. The first (5.1) is an ultimate extension (first announced in Lefschetz [g]) of a mapping theorem on the nerves of finite open coverings due to Alexandroff [AT,a]. The second (19.1) due to the author [d], describes a deformation-retraction of open subsets of the Hilbert parallelotope onto Euclidean complexes. This proposition has made it possible to free Borsuk's retraction theory from dimensional restrictions, and also to extend it in large measure to separable metric spaces.

§1. FUNDAMENTAL MAPPING THEOREM

1. Let $\mathfrak{R}$ be a topological space and $\mathfrak{U} = \{U_a\}$ an open covering of $\mathfrak{R}$. By an Euclidean, or a <u>geometric</u> nerve of $\mathfrak{U}$, we understand an Euclidean, or a geometric realization of nerve $\mathfrak{U}$. It is upon the geometric nerves that $\mathfrak{R}$ is mapped in the sequel.

Let Φ designate a geometric nerve of $\mathfrak{U}$. The simplexes of Φ shall be denoted by $U_1 \ldots U_j$ and their kernels by $[U_1 \ldots U_j]$ (AT,VII,1). In addition there will occur repeatedly the <u>strict intersection</u> of the sets $U_1, \ldots, U_j$, that is to say the set of the points which are on $U_1, \ldots, U_j$ but on no other sets of $\mathfrak{U}$, and it will be designated by $|[U_1 \ldots U_j]|$. Similar notations will be applied to any other covering.

The derived Φ', ... of Φ, will always be assumed barycentric. The vertices of Φ' are the centroids of the $U_i \ldots U_j$ and to be denoted by $(U_i \ldots U_j)_g$.

2. Let $\mathfrak{U}$ be a point-finite covering of $\mathfrak{R}$. An open covering $\mathfrak{U}'$ is said to be a <u>first derived</u> or merely a <u>derived</u> of $\mathfrak{U}$, whenever it is possible to label its sets $U'_{i\ldots j}$ under the following conditions:—

(2.1) there is at most a single set $U'_{i\ldots j}$ for each intersection $U_i \cap \ldots \cap U_j \neq \phi$ and the latter intersection contains $U'_{i\ldots j}$;

(2.2) the only non-vacuous intersections of sets of $\mathfrak{U}'$ are of the form $U'_{i\ldots j} \cap U'_{i\ldots j\ldots k} \cdots \cap U'_{i\ldots j\ldots k\ldots l}$.

From (2.1) follows at once that if $\mathfrak{U}'$ exists it is also point-finite. As for (2.2) it asserts in substance that in Φ:

$$U_i \ldots U_j \prec \ldots \prec U_i \ldots U_l.$$

Therefore, by (I, 4.g), if we choose the $(U_i \ldots U_j)_g$ as the vertices corresponding to the $U'_{i\ldots j}$ in constructing a geometric nerve Φ_1 for $\mathfrak{U}'$, then the simplexes of Φ_1 are merely elements of Φ'. Therefore Φ_1 is a closed subcomplex of Φ'.

When $\mathfrak{U}'$ has a first derived $\mathfrak{U}''$, the latter is called the <u>second</u> derived of $\mathfrak{U}$, etc. A point-finite open covering $\mathfrak{U}$, such that there exists a sequence $\mathfrak{U}^0 = \mathfrak{U}$, $\mathfrak{U}'$, $\mathfrak{U}''$, ... where $\mathfrak{U}^{(n+1)}$ is a derived of $\mathfrak{U}^{(n)}$, shall be called <u>analytic</u>, (Lefschetz [g]).

We notice the following properties:

(2.3) The derived, when it exists, need not be unique. (See end of 2.6).

(2.4) If A is a subset of $\mathfrak{R}$ and $\mathfrak{U} = \{U_i\}$ is a point-finite open covering of $\mathfrak{R}$ with a derived then $\mathfrak{B} = \{A \cap U_i\}$ is a similar covering for A. Therefore if $\mathfrak{U}$ is analytic so is $\mathfrak{B}$.

(2.5) Let $\mathfrak{R}$ be mapped continuously onto a topological space $\mathfrak{R}_1$ under a transformation T, and let $\mathfrak{W} = \{W_i\}$ be an open covering of $\mathfrak{R}_1$ and $\mathfrak{U} = \{T^{-1}W_i\}$, its inverse image in $\mathfrak{R}$. Then when $\mathfrak{W}$ is point-finite so is $\mathfrak{U}$, when $\mathfrak{W}$ has a derived so has $\mathfrak{U}$, and when $\mathfrak{W}$ is analytic so is $\mathfrak{U}$.

The first two properties are obvious and (2.5) is a consequence of the fact that T^{-1} preserves the mutual inclusions and intersections of the open sets.

(2.6) An example. Let K be any geometric complex and A_i its vertices. The open covering $\mathfrak{U} = \{|\text{St } A_i|\}$ of $|K|$ is clearly point-finite and we shall show that it is analytic. First the open covering $\mathfrak{U}'$ consisting of the stars of the vertices of the derived K' of K is a derived of $\mathfrak{U}$. For let $U_i = |\text{St } A_i|$, $\sigma = A_i \dots A_j \in K$, $\hat{\sigma}$ = the centroid of σ. We have then $U_i \cap \dots \cap U_j = |\text{St } \sigma|$. Let us set $U'_{i\dots j} = |\text{St } \hat{\sigma}|$ (star in K'). From the symbols of the elements of K' (II, 4.9) we verify readily: $|\text{St } \hat{\sigma}| \subset |\text{St } \sigma|$, or $U'_{i\dots j} \subset U_i \cap \dots \cap U_j$. Thus the basic property (2.1) is fulfilled by $\mathfrak{U}'$. Regarding (2.2) if several sets U' intersect, their intersection is an $|\text{St } \sigma'|$, $\sigma' \in K'$. Therefore the vertices of K' which are the centers of the stars U' of K' are those of a simplex of K'. Hence (loc. cit.) the sets U' in question must satisfy (2.2). It follows that $\mathfrak{U}'$ is a derived of $\mathfrak{U}$.

Evidently the stars of the vertices of the successive derived $K^{(n)}$ of K form a sequence of derived coverings of $\mathfrak{U}$. Therefore $\mathfrak{U}$ is analytic.

We may observe that if we carry out the construction of the derived with centroids replaced by other points on the respective simplexes, we obtain a derived $\mathfrak{U}_1' \neq \mathfrak{U}'$. Similarly of course for the other derived. This shows clearly that the $\mathfrak{U}^{(n)}$ need not be unique.

3. We shall now assume that we have an analytic open covering $\mathfrak{U}$ of $\mathfrak{R}$ and determine a mapping associated with $\mathfrak{U}$.

The notations being as before we have then a sequence of derived $\mathfrak{U}^0 = \mathfrak{U}$, $\mathfrak{U}'$, ... Let Φ_p be a geometric nerve for $\mathfrak{U}^{(p)}$, with $\Phi_0 = \Phi$ and Φ_{p+1} a subcomplex of Φ_p'.

The following convenient notations will be used. Whenever we do not wish to indicate explicitly the relation of the sets of $\mathfrak{U}^{(p)}$ to those of its predecessors we shall denote them by U_i^p, with $U_i^0 = U_i$.

Consider now the transformation of the points of $\mathfrak{R}$ into the closed simplexes of Φ defined by

$$(3.1) \qquad \tau = (\tau\mathfrak{U}): \ |[U_i \dots U_j]| \longrightarrow \overline{U_i \dots U_j}.$$

It is uniquely defined for every point-finite open covering $\mathfrak{U}$, whether analytic or otherwise. However when $\mathfrak{U}$ is analytic we have similar mappings $\tau_p = \tau(\mathfrak{U}^{(p)})$ for the derived coverings. For a given point x of $\mathfrak{R}$ let

$$(3.2) \qquad \tau x = \overline{U_i \dots U_j}, \ x \in |[U_i \dots U_j]|;$$

$$(3.3) \qquad \tau_1 x = \overline{U'_{h\dots k} \dots U'_{h\dots k\dots l}},$$
$$x \in |[U'_{h\dots k} \dots U'_{h\dots k\dots l}]|.$$

From (3.2) follows that x is on the sets $U_i, \dots, U_j$ and on no others, and from (3.3) that it is on $U'_{h\dots k\dots l}$, and hence by (2.2) that it is on $U_h \cap \dots \cap U_k \cap \dots U_l$.

Therefore the indices $h, \dots, l$ are among the indices $i, \dots, j$, or:

(3.4) $\tau x \supset \overline{U_h \dots U_1 \dots U_m}$.

Let ζ be the simplex of Φ such that $\bar{\zeta} = \tau x$. It is a consequence of (3.4) that $\tau_1 x$ is a simplex of the first derived $(\mathrm{Cl}\ \zeta)'$. For convenience let the latter be written $(\tau x)'$ (as if τx were a complex and not merely a point-set), and similarly for any $\tau_p x$. Thus by (3.4) $\tau_1 x$ is a closed simplex of $(\tau x)'$ or:

(3.5) $$\tau_1 x \subset (\tau x)'.$$

The same relation for τ_{p+1} is

(3.6) $$\tau_{p+1} x \subset (\tau_p x)',$$

and hence

(3.7) $$\tau_p x \subset (\tau x)^{(p)}.$$

Explicitly: $\tau_p x$ is a simplex of the $p^{\underline{th}}$ derived of τx.

4. By (I,3.8) $y = \bigcap \tau_p x$ is a single point of x. Let T denote the transformation $\mathfrak{R} \longrightarrow \Phi$ such that $Tx = y$. We shall examine the properties of T and in particular prove that it is continuous.

First of all $Tx \subset \tau x$, and hence

(4.1) $$T|[U_i \dots U_j]| \subset \overline{U_i \dots U_j}.$$

Or explicitly: T sends every strict intersection of sets of $\mathfrak{U}$ into the closed simplex which represents the intersection in the nerve Φ of $\mathfrak{U}$. If we write (4.1) for $\mathfrak{U}^{(p)}$, and notice that

$$U_i^p \cap \dots \cap U_j^p = \bigcup |[U_i^p \dots U_j^p U_h^p \dots U_k^p]|,$$

we have at once

(4.2) $$TU_i^p \cap \dots \cap U_j^p \subset \overline{|\mathrm{St}\ U_i^p \dots U_j^p|}.$$

Let now $x \in |[U_h^p \dots U_k^p]|$, $Tx \in |\mathrm{St}\ U_i^p|$. We have then also $Tx \in \tau_p x = U_h^p \dots U_k^p$. Therefore the latter has a face in $\mathrm{St}\ U_i^p$, and hence i is among the indices $h, \dots, k$, or finally $x \in U_i^p$. Therefore

(4.3) $$T^{-1}\ |\mathrm{St}\ U_i^p| \subset U_i^p.$$

From this follows:

(4.4) $$T^{-1} |\mathrm{St}\ U_1^p \dots U_j^p| \subset \cdot U_1^p \cap \dots \cap U_j^p.$$

5. We pass to the proof that T is continuous. Let $y = Tx$, and let V be a neighborhood of y in Φ, and σ_p the simplex of $\Phi^{(p)}$ containing y. The stars of $\Phi^{(p)}$ containing y are those of the faces of σ_p and they all contain $\mathrm{St}\ \sigma_p$. Therefore $\{|\mathrm{St}\ \sigma_p|\}$ is a base for y. It follows from this, and since Φ is a metric space, that for some p : $\overline{|\mathrm{St}\ \sigma_{p-1}|} \subset V$. Let $x \in U_1^p \cap \dots \cap U_j^p = U$. From (4.2) we conclude that $y \subset |\mathrm{St}\ U_1^p \dots U_j^p|$. Therefore the latter contains σ_p, and hence finally $U_1^p \dots U_j^p \subset \mathrm{Cl}\ \mathrm{St}\ \sigma_p$. Therefore $|\mathrm{St}\ U_1^p \dots U_j^p| \subset |\mathrm{St}\ \mathrm{Cl}\ \mathrm{St}\ \sigma_p|$ (star in $\Phi^{(p)}$) $\subset |\overline{\mathrm{St}\ \sigma_{p-1}}|$ (star in $\Phi^{(p-1)}$) $\subset V$. In other words x has the neighborhood U such that $TU \subset V$. Therefore T is continuous.

A mapping T: $\mathfrak{R} \longrightarrow \Phi$ for which (4.1) holds shall be called <u>canonical relative</u> to $\mathfrak{U}$. We have thus shown that T is canonical relative to $\mathfrak{U}$, and similarly of course relative to every $\mathfrak{U}^{(p)}$.

Conversely let T be a mapping of a topological space $\mathfrak{R}$ into a geometric complex Φ and let $\{A_i\}$ be the vertices of Φ, and $W_i = |\mathrm{St}\ A_i|$, $U_i = T^{-1}\ |\mathrm{St}\ A_i|$, $\mathfrak{U} = \{U_i\}$, $\mathfrak{W} = \{W_i\}$. As we have seen $\mathfrak{W}$ is analytic (2.6), and hence so is $\mathfrak{U}$ (2.5). Moreover $T|[U_i \dots U_j]| \subset |\mathrm{St}\ A_i \cap \dots \cap \mathrm{St}\ A_j| = A_i \dots A_j = U_i \dots U_j$. Therefore T is canonical relative to $\mathfrak{U}$.

We may even go a step farther. Suppose merely that $\mathfrak{U}$ is an open covering with a geometric nerve Φ and that there exists a mapping T: $\mathfrak{R} \longrightarrow \Phi$ which is canonical relative to $\mathfrak{U}$. Then we shall show that $\mathfrak{U}$ is analytic. Consider in fact $W_i = T^{-1}|\ \mathrm{St}\ U_i|, \mathfrak{W} = \{W_i\}$. By (4.3) for $p = 0$: $\mathfrak{W} \succ \mathfrak{U}$. Moreover $\mathfrak{W}$ is analytic (2.6). Let $\mathfrak{W}'$ be a derived of $\mathfrak{W}$: $\mathfrak{W}' = \{W'_{i \dots j}\}$, the labels being in accordance with (2.1, 2.2). Corresponding to $U_i \cap \ldots \cap U_j \neq \Phi$ there is a single $W'_{i \ldots j} \subset W_i \ldots W_j \subset U_i \ldots U_j$ (the last step by (4.4) for $p = 0$). Therefore (2.1) is fulfilled by $\mathfrak{W}'$ relative to $\mathfrak{U}$. Since (2.2) holds relative to $\mathfrak{W}$ it does likewise relative to $\mathfrak{U}$. Therefore $\mathfrak{W}'$ is a derived of $\mathfrak{U}$ and since $\mathfrak{W}'$ is analytic so is $\mathfrak{U}$.

To sum up then we have proved the

(5.1) FUNDAMENTAL MAPPING THEOREM. A n.a.s.c. in order that a topological space $\mathfrak{R}$ possess a canonical mapping relative to an open covering $\mathfrak{U}$ of $\mathfrak{R}$, is that $\mathfrak{U}$ be analytic. Furthermore every continuous mapping of $\mathfrak{R}$ on a geometric complex is a canonical mapping relative to some open covering of $\mathfrak{R}$ (Lefschetz [g]).

6. Before proceeding it is necessary to consider certain mappings of geometric complexes into one another. Let then $K = \{\sigma\}$ be a geometric complex, $[K]$ its simplicial antecedent (I,5), $\hat{\sigma}$ the centroid of σ. Similar notations are applied to a second complex $L = \{\zeta\}$ and to the derived of K, L.

Suppose now that we have a simplicial set-transformation $\tau: [K] \longrightarrow [L]$. We define a new simplicial set-transformation $\tau': [K'] \longrightarrow [L']$ the first <u>derived</u> of τ, by the condition that if $\tau\sigma = \zeta$ then $\tau'\hat{\sigma} = \hat{\zeta}$. To show that τ' is in fact simplicial notice that any simplex of $[K']$ is of the form $\hat{\sigma}_i \ldots \hat{\sigma}_j$, where $\sigma_i < \ldots < \sigma_j$. Now if $\tau\sigma_i = \zeta_i$ likewise $\zeta_i < \ldots < \zeta_j$ and so $\tau'\hat{\sigma}_i \ldots \hat{\sigma}_j = \hat{\zeta}_i \ldots \hat{\zeta}_j$ is a simplex of $[L']$. We define all the derived τ'', τ''', ... by $\tau^{(n+1)} = (\tau^n)'$.

Given any $x \in |K|$, denote by $\sigma_n(x)$ the simplex of $K^{(n)}$ containing x. We have at once $\tau^{(n+1)}\,\bar{\sigma}_{n+1}(x)$

$\subset \tau^{(n)} \bar{\sigma}_n(x)$. Therefore $y = \bigcap \tau^n \bar{\sigma}_n(x)$ is a point of $|L|$ (I, 3.8). Let $\tau^{(\omega)}$ denote the (point-set) transformation $|K| \longrightarrow |L|$ such that $\tau^{(\omega)}x = y$. It is simplicial in the sense that $\tau^{(\omega)}\sigma$ is a ζ. Moreover let $x_p \longrightarrow x$ on $|K|$. For p above a certain value $x_n \in |\mathrm{St}\ \sigma_n(x)|$, hence $\tau^{(\omega)}x_n \in |\mathrm{St}\ \sigma^{(n)}x|$. Since the diameter of the star < 2 mesh $L^{(n)}$ which $\longrightarrow 0$, $\tau^{(\omega)}x_n \longrightarrow \tau^{(\omega)}x$. Therefore $\tau^{(\omega)}$ is continuous and so:

(6.1) $\tau^{(\omega)}$ is a continuous mapping $|K| \longrightarrow |L|$ inducing the simplicial set-transformations:

$\tau^{(n)}$: $[K^n] \longrightarrow [L^{(n)}]$ $(\tau^{(0)} = \tau)$.

An immediate consequence is

$$(6.2) \qquad \tau\sigma = \zeta \Longrightarrow \tau^{(\omega)}\bar{\sigma} = \bar{\zeta}.$$

7. Returning to our main argument we shall introduce, for later purposes, a generalization of (5.1). Given the same topological space $\mathfrak{R}$, and the point-finite open covering $\mathfrak{U} = \{U_i\}$ let there exist a refinement $\mathfrak{U}'$ whose sets may be labelled $U^l_{i \ldots j}$, and so that:

(7.1) There may exist one or more sets $U^l_{i \ldots j}$ for each intersection $U_i \cap \ldots \cap U_j \neq \emptyset$ and $U^l_{i \ldots j} \subset U_i \cap \ldots \cap U_j$;

(7.2) The only non-void intersections of sets of $\mathfrak{U}'$ are of the form

$$U^l_{i \ldots j} \cap U^m_{i \ldots j \ldots h} \cap \ldots \cap U^n_{i \ldots j \ldots h \ldots k}.$$

When $\mathfrak{U}'$ exists it is called a <u>generalized derived</u> (= g.d.) of $\mathfrak{U}$. If there exists a <u>generalized derived sequence</u> $\mathfrak{U}^0 = \mathfrak{U}, \mathfrak{U}', \mathfrak{U}'', \ldots$, i.e. a sequence in which each covering is a generalized derived of its predecessor, $\mathfrak{U}$ is said to be a generalized analytic covering

(= g.a.c.).

From (7.1), (7.2) follow immediately:

(7.3) A refinement of a generalized derived or generalized derived of a refinement is a generalized derived.

Furthermore

(7.4) If $\tilde{\mathfrak{U}}'$ is a generalized derived of $\mathfrak{U}$ then there is a derived $\mathfrak{U}'$ such that $\tilde{\mathfrak{U}}' \succ \mathfrak{U}' \succ \mathfrak{U}$.

For if $U^l_{i \dots j}$ are the sets of $\tilde{\mathfrak{U}}$ and we set $U_{i \dots j} = \bigcup_l U^l_{i \dots j}$, then $\mathfrak{U}' = \{U_{i \dots j}\}$ behaves as required.

(7.5) A derived is also a generalized derived and an analytic covering is also generalized analytic. (Obvious)

If we combine (7.3) and (7.4) we find at once that

(7.6) If $\mathfrak{U}$ is generalized analytic it is analytic.

In other words generalized derivation is only of interest as a property of a certain sequence $\{\mathfrak{U}^{(n)}\}$, but yields nothing new as regards the coverings.

(7.7) Let $\mathfrak{U}$ be point-finite and let it have a derived $\mathfrak{U}'$. Then if $\mathfrak{U}'_1$ is a point-finite refinement of $\mathfrak{U}'$, it is a generalized derived of $\mathfrak{U}$.

Let $\mathfrak{U}, \mathfrak{U}'$ be as in (1) and let the sets of $\mathfrak{U}'_1$

receive labels $U^1_{i...j}$ such that $U^1_{i...j}$ is a definite set of $\mathfrak{U}'$ which $\supset U^1_{i...j}$. Then property (7.1) holds and (7.2) is a consequence of (2.2). Therefore $\mathfrak{U}'_1$ is a generalized derived of $\mathfrak{U}$.

Let now $\mathfrak{U}$ be a g.a.c. with $\mathfrak{U}^o = \mathfrak{U}, \mathfrak{U}', \ldots$, as above and Φ_p as geometric nerve for $\mathfrak{U}^{(p)}$. This time the vertex-mapping of the simplicial antecedent of Φ_1 into the one of Φ' sending the vertex $U^1_{i...j}$ into $(U^1_i \ldots U^1_j)_g$ (the centroid of $U^1_i \ldots U^1_j$), merely determines a simplicial mapping of the antecedents into one another. Let $\theta_o = \eta^{(\omega)}$ (in the sense of 6). There is a similar mapping θ_p: $\Phi_{p+1} \longrightarrow \Phi'_p$ for every p. The aggregate $\theta = \{\theta_p\}$ may be considered as a unique projection operating on $\Phi_1, \Phi_2, \ldots$ and coincident with θ_p on Φ_{p+1}.

We now introduce the mappings τ_p of (3), prove as in (3) the relation

$$(7.9) \qquad \theta\, \tau_1 x \subset (\tau x)',$$

analogous to (3.5), and by obvious repetition the more general

$$(7.10) \qquad \theta^p\, \tau_{n+p} x \subset (\theta^{p-1}\, \tau_{n+p-1} x)'.$$

As loc. cit. $y_n = \bigcap \tau_{n+p} x$ is a single point of Φ_n and T_n: $x \longrightarrow y_n$ is a mapping $\mathfrak{R} \longrightarrow \Phi_n$.

If we denote by U^p_i the sets of $\mathfrak{U}^{(p)}$ we prove again (4.1, 4.3) as in (4) for each T_n. More explicitly

$$(7.11) \qquad T_n | [U^n_i \ldots U^n_j] | \subset \overline{U^n_i \ldots U^n_j},$$

$$(7.12) \quad T_n^{-1} \theta^p | \mathrm{St}\ U^{n+p}_i \ldots U^{n+p}_j | \subset U^{n+p}_i \ldots U^{n+p}_j.$$

From this we conclude as before that T_n is canonical relative to $\mathfrak{U}^{(n)}$. Finally from the definition of T_n follows $\theta y_{n+1} = y_n$ or $\theta T_{n+1} = T_n$. We have therefore:

(7.13) Let the topological space $\mathfrak{R}$ admit an analytic covering $\mathfrak{U}$ with associated generalized derived coverings $\mathfrak{U}'$, $\mathfrak{U}''$, ..., and geometric nerves Φ_p with projection θ: $\Phi_{n+1} \longrightarrow \Phi'_n$. Then there exists a sequence of canonical mappings T_n: $\mathfrak{R} \longrightarrow \Phi_n$ such that $\theta T_{n+1} = T_n$.

§2. APPLICATION TO NORMAL AND TYCHONOFF SPACES

8. The simplest possible non-trivial open covering of a topological space $\mathfrak{R}$ is a <u>binary</u> covering $\mathfrak{U} = (U_1, U_2)$ consisting of two proper open subsets. A geometric nerve is here merely a segment l: $0 \leq t \leq 1$. We may assume that the end-points $0, 1$ represent respectively U_1 and U_2, and the interval represents $U_1 \cap U_2$. Since U_1, U_2 are proper sub-sets of $\mathfrak{R}$, $F_1 = |[U_1]|$ and $F_2 = |[U_2]|$ are non-void disjoint closed sets. Analyticity implies that there exists a canonical mapping relative to $\mathfrak{U}$. We may consider it as a continuous function $f(x)$ on $\mathfrak{R}$ to l, and since it is canonical it maps F_1 on $t = 0$, F_2 on $t = 1$. In other words f is a <u>characteristic function</u> for the two sets. Thus the analyticity of $\mathfrak{R}$ implies the existence of a characteristic function for F_1, F_2 or equivalently that the normality condition is fulfilled by the two sets (AT,I,34.1). Thus normality and analyticity are closely related. We prove in fact:

(8.1) THEOREM. Every locally finite open covering $\mathfrak{U}$ of a normal space $\mathfrak{R}$ is analytic.

Let $\mathfrak{U} = \{U_i\}$ and let Φ be a geometric nerve for $\mathfrak{U}$. By(AT,I,33.4) $\mathfrak{U}$ may be shrunk to a covering $\mathfrak{V} = \{V_i\}$ where $V_i \subset U_i$. Since $\mathfrak{R}$ is normal the disjoint closed sets $\mathfrak{R} - U_i$, $\overline{V}_i$ have a characteristic function $f_i(x)$. Since $\mathfrak{V}$ is a covering at least one

$f_i = 1$ at x, and since $\mathfrak{U}$ is point-finite at most a finite number of f_i are $\neq 0$ at x. Therefore

(8.2) $$y_i = \frac{f_i(x)}{\sum f_j(x)}$$

is a single-valued function of x. If $U_1, \ldots, U_j$ are the sets of $\mathfrak{U}$ containing x, we have $y_h = 0$ for $h \neq 1, \ldots, j$, and clearly

(8.3) $$y_i \geqq 0; \quad \sum y_i = 1.$$

Therefore the y_i are the barycentric coordinates of a point y on the simplex $U_1 \ldots U_j$ of Φ. It follows that the transformation $T: x \longrightarrow y$, except perhaps for continuity, behaves as required. To prove continuity let Φ be assigned the Euclidean metric

(8.4) $$d(y,y') = (\sum (y_i - y_i')^2)^{1/2}.$$

Since Φ is locally finite (8.4) defines a topology equivalent to that of Φ (I, 9.11). Let now $W = U_1 \cap \ldots \cap U_j$, and let $U_1, \ldots, U_j, \ldots, U_k$ be the sets of $\mathfrak{U}$ which meet one of the sets $U_1, \ldots, U_j$. The only barycentric coordinates that may be $\neq 0$ for any $x \in W$ are $y_1, \ldots, y_k$. Therefore on W

$$y_h = \frac{f_h(x)}{f_1(x) + \ldots + f_k(x)},$$

and so y_h is a continuous function of x on W and hence on $\mathfrak{R}$ itself. From this follows immediately that $d(Tx,Tx')$ is continuous in (x,x'), and then at once that T is continuous (I, 9.6). This proves (8.1).

(8.5) COROLLARY. Every locally finite open covering of a metric space is analytic.

(8.6) COROLLARY. For a topological space normality is equivalent to the condition that every finite open covering be analytic.

Necessity has just been proved. Sufficiency follows from the fact that the condition implies that all the binary open coverings are analytic, which as we have seen implies normality.

(8.7) THEOREM. A n.a.s.c. for a normal space $\mathfrak{R}$ to be n dimensional, is that n be the least integer with the following property: given an arbitrary finite open covering $\mathfrak{U}$ of $\mathfrak{R}$ there exists a continuous mapping of $\mathfrak{R}$ into a finite polyhedron $|\Phi|$ whose dimension $\leq n$, such that if A_i are the vertices of Φ then $\{|T^{-1}StA_i|\}$ refines $\mathfrak{U}$.

(8.8) More generally the same property holds with $\mathfrak{U}$ merely any covering of a collection cofinal in the totality of all the finite open coverings.

(8.9) THEOREM. A n.a.s.c for a compactum $\mathfrak{R}$ to be n dimensional is that n be the least integer such that whatever $\epsilon > 0$, Φ possesses an ϵ-mapping into a finite polyhedron whose dimension $\leq n$. (Alexandroff, AT [a]).

Alexandroff's result is the prototype of all such mapping theorems. It is clear that (8.8) is a consequence of (8.7). As for (8.9) take a collection of finite open coverings $\{\mathfrak{U}_n\}$ whose meshes $\epsilon_n \longrightarrow 0$. Since $\{\mathfrak{U}_n\}$ is cofinal in the totality of all the finite open coverings (8.8) applies and so (8.9) will hold for the value ϵ_n of ϵ, and hence for every ϵ.

There remains then to prove (8.7). If $\dim \mathfrak{R} = n$, $\mathfrak{U}$ has a finite refinement $\mathfrak{B}$ whose order $\leq n$. By

(8.1) $\mathfrak{B}$ is analytic. Hence if Φ is a geometric nerve of $\mathfrak{B}$, the canonical mapping $T\colon \mathfrak{R} \longrightarrow \Phi$, relative to $\mathfrak{B}$ behaves as required. Conversely if T exists with the properties described in (8.7), and since order $\{T^{-1}|\mathrm{St}A_1|\} \leq n$, we have $\dim \mathfrak{R} \leq n$. Since n is the least integer with this property, actually $\dim \mathfrak{R} = n$, else by the necessity proof $\mathfrak{R}$ could be mapped in the asserted way on a $|\Phi|$ whose dimension $< n$.

9. We shall now prove a theorem which ties up Tychonoff spaces with analytic coverings.

> (9.1) THEOREM. A n.a.s.c. for a T_0 space to be a Tychonoff space is the existence of a family of finite analytic coverings $\{\mathfrak{U}_\alpha\}$ such that the corresponding canonical mappings $\kappa = \{T_\alpha\}$ form a basic class (AT,I,35.5).

Suppose $\mathfrak{R}$ to be a Tychonoff space. By (AT,I,35.8), it has a topological image R in a compact parallelotope $P = \mathsf{P} l_\alpha$, l_α a segment. By (5.1) the projections $\pi_\alpha : P \longrightarrow l_\alpha$ form a basic class κ. Hence the induced mappings $T_\alpha : R \longrightarrow l_\alpha$ form a basic class for R. This proves necessity.

Suppose now that the condition of (9.1) holds and let Φ_α be a geometric nerve for $\mathfrak{U}_\alpha$, and $\Phi = \mathsf{P}|\Phi_\alpha|$. We may assume that Φ_α is in an Euclidean parallelotope P^{n_α} where $n_\alpha = 2\dim \Phi_\alpha + 1$. Hence Φ may be mapped topologically in $\mathsf{P} P^{n_\alpha}$, which is a compact parallelotope. Therefore to prove sufficiency we merely need to show that $\mathfrak{R}$ may be mapped topologically on Φ.

Given $x \in \mathfrak{R}$ let $y_\alpha = T_\alpha x$ and let $y = \{y_\alpha\} \in \Phi$. Define a mapping $T\colon \mathfrak{R} \longrightarrow \Phi$ by $Tx = y$. Our aim will be achieved if we can prove T topological.

Since $\mathfrak{R}$ is a T_0 space, by (AT,I,35.6), κ is a separating class. Therefore $(x \neq x') \Longrightarrow (T_\alpha x \neq T_\alpha x'$ for some $\alpha) \Longrightarrow (Tx \neq Tx')$. Hence T^{-1} is single-valued like T, and so T os one-one.

Since κ is a basic class T maps into one another the elements of two subbases for $\mathfrak{R}$ and Φ, and so it is bicontinuous. Therefore T is topological and the theorem is proved.

§3. COMPACT IMBEDDING OF SEPARABLE METRIC SPACES.

10. We have already seen (AT,I,46.3) that a separable metric space $\mathfrak{R}$ may always be topologically immersed in the Hilbert parallelotope P^{ω}. When $\dim \mathfrak{R} = \infty$ this is about as far as we may go. In point of fact if $\mathfrak{R}$ is already assumed in P^{ω} then $\overline{\mathfrak{R}}$ is a compactum in which $\mathfrak{R}$ is immersed topologically and densely, and this is a mild and sometimes useful complement. There exists however the following more accurate result valid for all dimensions:

(10.1) THEOREM. A separable metric space $\mathfrak{R}$ may always be immersed topologically and densely in a compactum $\mathfrak{S}$ of the same dimension. (Hurewicz [a]).

11. For the proof we shall require the following properties:

(11.1) If A is a totally bounded separable metric space whose dimension $\leqq n$, then whatever $\epsilon > 0$ there exists a finite ϵ open covering $\mathfrak{U}$ of A such that $\overline{\mathfrak{U}}$ (and hence also $\mathfrak{U}$) is of order $\leqq n$.

Let $\mathfrak{V}$ be a finite ϵ open covering. Since $\dim A \leqq n$, $\mathfrak{V}$ has a refinement $\mathfrak{W} = \{W_1, \ldots, W_r\}$ whose order $\leqq n$. Since A is normal $\mathfrak{W}$ may be shrunk to $\mathfrak{U} = \{U_1, \ldots, U_r\}$ such that $\overline{U}_i \subset W_i$. Therefore mesh $\mathfrak{U} \leq$ mesh $\mathfrak{W} \leq$ mesh $\mathfrak{V} < \epsilon$, and order $\mathfrak{U} \leq$ order $\overline{\mathfrak{U}} \leq$ order $\mathfrak{W} \leq n$, which proves (11.1).

(11.2) If a compactum (or more generally any normal space) A has finite closed coverings of arbitrarily small mesh whose order $\leq n$, then $\dim A \leq n$.

Let $\mathfrak{U} = \{U_i\}$ be any finite ϵ open covering. By assumption there is a finite closed covering $\mathfrak{F} = \{F_j\}$ whose order $\leq n$ and whose mesh $<$ Lebesgue number $\mathfrak{U}$. As a consequence $\mathfrak{F} \succ \mathfrak{U}$. By (AT,I,33.5) we can find for each F_j an open set V_j such that $F_j \subset V_j$ and that $F_j \longrightarrow V_j$ is a similitude of $\mathfrak{F}$ with $\mathfrak{V} = \{V_j\}$. Select for each F_j an open set $U_{i(j)} \supset F_j$, and replace V_j by $W_j = V_j \cap U_{i(j)}$. Evidently $\mathfrak{W} = \{W_j\}$ behaves like $\mathfrak{V}$. Hence like $\mathfrak{V}$ it is of order $\leq n$, and since $\mathfrak{W} \succ \mathfrak{U}$, $\dim A \leq n$.

Finally from general dimension theory (see Hurewicz-Wallman, AT [H-W] p. 26) we borrow without proof:

(11.3) If A, B are separable metric and $A \subset B$ then $\dim A \leq \dim B$.

12. We are now ready for the proof of (10.1). Let the Hilbert parallelotope P^ω be referred to the coordinates $x_1, x_2, \ldots$, where x_n has for range $l_n : 0 \leq x_n \leq 1$. Thus P^ω is the metric product $\mathsf{P} l_n$. We also introduce $\mathsf{P}'^\omega = \mathsf{P} l_{2n+1}$, $\mathsf{P}''^\omega = \mathsf{P} l_{2n}$, and both are Hilbert parallelotopes. We shall also require the finite parallelotope $P_m = l_2 \times l_4 \times \cdots \times l_{2m}$.

Since P'^ω is a Hilbert parallelotope, $\mathcal{R}$ has a topological image R_o in P'^ω. Choose now a sequence of positive constants $\epsilon_m \longrightarrow 0$ which are to be the meshes of certain coverings. By (11.1) we may find a finite ϵ_o open covering $\mathfrak{U}_o$ of R_o whose order $\leq n$. Let

$\mathfrak{U}_o = \{U_{oi}\}$ and let F_{oh} $(h = 1, 2, \ldots, s)$ denote the intersections of the sets $\overline{U}_{oi}$. It may well happen that $F_{oh} \cap F_{oj} = \emptyset$ and yet $d(F_{oh}, F_{oj}) = 0$. Corresponding to such a pair however there will exist a characteristic function $f(x)$. To be precise we suppose $h < j$ and $f = 0$ on F_{oh}, $f = 1$ on F_{oj}. Let all the characteristic functions thus obtained be denoted in some order by $f_2(x), \ldots, f_{2n_1}(x)$. If x is the point $(x_1, 0, x_3, \ldots)$ of R_o and

$$s_1 x = (x_1, \frac{f_2(x)}{2}, \ldots, \frac{f_{2n_1}(x)}{2n_1}, x_{2n+1}, 0, \ldots),$$

then s_1 defines a mapping $R_o \longrightarrow P'^{\omega} \times P_{n_1}$ which by (AT,I,12.4) maps R_o topologically onto a subset $R_1 = s_1 R_o$ of the product. It is clear that $\mathfrak{B}_1 = s_1 \mathfrak{U}_o = \{s_1 U_{oi}\}$ is a finite open covering of R_1 whose order $\leq n$ and which has the following basic property: if F, F' are any two intersections of sets of $\overline{\mathfrak{B}}_1$ (closures in R_1) which do not meet, then $d(F, F') > 0$. Let this property be denoted by (α).

Suppose that we have obtained similarly in a certain $P'^{\omega} \times P_{n_k}$ a topological image R_k of $\mathfrak{R}$ having a finite open covering $\mathfrak{B}_k$ whose order $\leq n$ and which has the property (α). We construct next a finite open covering $\mathfrak{U}_k$ of R_k whose mesh $\leq \epsilon_k$ and such that order $\mathfrak{U}_k \leq n$. Then proceeding as above by means of certain characteristic functions $f_{2(n_k+1)}, \cdots, f_{2n_{k+1}}$ we define the topological mapping $s_{k+1} : R_k \longrightarrow R_{k+1}$, where R_{k+1} is in $P'^{\omega} \times P_{n_{k+1}}$, and the new coordinates of $s_{k+1}x^k$ $(x^k \in R^k)$ which are $\neq 0$ will be:

$$(12.1) \quad x_{2\nu} = \frac{f_{2\nu}(x^k)}{2\nu}, \quad \nu = n_k + 1, \ldots, n_{k+1}.$$

If we set $s_k s_{k-1} \ldots s_1 = \sigma_k$, we have $x^k = \sigma_k x$, and

σ_k is a topological mapping $R_o \longrightarrow R_k$. From its definition follows that if π_k is the projection $P'^{\omega} \times P_{n_k} \longrightarrow P'$ then $\sigma_k^{-1} = \pi_k | R_k$. If we set

$$(12.2) \qquad \frac{f_{2\nu}(\sigma_k x)}{2\nu} = \phi_{2\nu}(x), \nu = n_k+1, \ldots, n_{k+1},$$

we may replace (12.1) by

$$(12.3) \qquad x_{2\nu} = \phi_{2\nu}(x), \nu = n_k+1, \ldots, n_{k+1},$$

where $\phi_{2\nu}$ is continuous and single-valued on R and has for range $0 - \frac{1}{2\nu}$. Finally our construction is such that $\mathfrak{V}_{k+1} = s_{k+1}\mathfrak{U}_k$ is a finite open covering of R_{k+1} whose order $\leq n$ and which has the property (α).

Let now $x = (x_1, 0, x_3, 0, \ldots) \in R_o$ and define $\sigma x = (x_1, x_2, \ldots), x_{2\nu} = \phi_{2\nu}(x)$. Since $x \longrightarrow (0, x_2, 0, x_4, \ldots)$ defines a mapping $R \longrightarrow P''^{\omega}$, by (AT,I,12.4)$\sigma$ maps R topologically onto a subset R^* of P^{ω}. Let τ_k denote the projection $P^{\omega} \longrightarrow P'^{\omega} \times P_{n_k}$. We have $\tau_k | R^* = \theta_k = \sigma_k \sigma^{-1}$, so θ_k is topological. If we set $\mathfrak{W}_k = \theta_k^{-1}\mathfrak{V}_k, \mathfrak{W}_k$ is a finite open covering of R^* with the following properties:

(a) order $\overline{\mathfrak{W}}_k \leq n$;

(b) if F, F' are intersections of sets of $\mathfrak{W}_k$ and $F \cap F' = \phi$ then $d(F,F') > 0$ (property α).

(c) mesh $\mathfrak{W}_k \rightarrow 0$.

Property (a) is a consequence of the same for $\mathfrak{V}_k$. Regarding (b) when F, F' are as stated one of the coordinates $\phi_{2\nu}$ of (12.2) vanishes on one of the two sets and has the value $\frac{1}{2\nu}$ on the other so that $d(F,F') \geq \frac{1}{2\nu}$. As for (c) denote by $\rho^2(s)$ the remainder after s terms of the convergent series $\sum s^{-2}$, so that $\rho(s) \longrightarrow 0$. From the definition of θ_k follows that if $x^* \in R^*$ then $d(x^*, \theta_k x^*) \leq \rho(2n-2) \longrightarrow 0$. Therefore mesh $\mathfrak{W}_k \leq 4n_k\rho + \epsilon_k \longrightarrow 0$ also and this is (c).

Let now S denote the closure of R^* in P^{ω} and if $\mathfrak{W}^k = \{W_{k1}, \ldots, W_{kr}\}$ let F_{ki} denote the closure of W_{ki} in P^{ω}. Then $\mathfrak{F}_k = \{F_{k1}, \ldots, F_{kr}\}$ is a finite closed covering of S. It is a consequence of (b), (c) that $\mathfrak{F}_k$ has the same order and also the same mesh as $\mathfrak{M}_k$. Therefore S possesses finite closed coverings of arbitrarily small mesh and whose order $\leqq$ n. Hence dim S $\leqq$ n (11.3). Since $R^* \subset S$, by (11.4): dim R^* = n $\leqq$ dim S. Hence finally dim S = n. Since R^* is a topological image of $\mathfrak{R}$ and is dense in S, (10.1) is completely proved.

§4. TOPOLOGICAL IMBEDDING IN EUCLIDEAN SPACES.

13. We shall now take up an imbedding theorem surmised in its general form by Menger and first proved by Nöbiling [a]. See also Lefschetz [a], Pontrjagin-Tolstowa [a], Hurewicz [d].

> (13.1) THEOREM. Every compactum $\mathfrak{R}$ of finite dimension n may be topologically imbedded in an Euclidean space of dimension $\leqq$ 2n+1.

Combining (13.1) with (12.1) we also have:

> (13.2) The preceding theorem holds likewise for separable metric spaces.

And as a noteworthy special case, first proved by Sierpinski:

> (13.3) Every separable metric zero-dimensional space is topologically equivalent to a subset of the real line.

14. Let $\mathfrak{U}$ be any finite open covering of $\mathfrak{R}$. Since $\mathfrak{R}$ is metric it is normal. Hence $\mathfrak{U}$ is analytic and therefore it has a derived $\mathfrak{U}'$. Since dim $\mathfrak{R}$ = n, $\mathfrak{U}'$ has a refinement $\mathfrak{V}$ of mesh less than an assigned ϵ

and whose order $\leq$ n. Referring to (7.3), $\mathfrak{B}$ is a generalized derived of $\mathfrak{U}$. Therefore $\mathfrak{U}$ has a generalized derived whose order $\leq$ n and whose mesh is arbitrarily small. It follows that we can find a sequence of generalized derived $\mathfrak{U}^o = \mathfrak{U}$, $\mathfrak{U}'$, $\mathfrak{U}''$, ... such that ϵ_p = mesh $\mathfrak{U}^{(p)} \longrightarrow 0$. As a consequence if Φ_p is a geometric nerve of $\mathfrak{U}^{(p)}$ we have by (7.13) the canonical mappings T_n: $\mathfrak{R} \longrightarrow |\Phi_p|$ such that if θ is the projection $|\Phi_{p+1}| \longrightarrow |\Phi_p'|$ of (7) then $\theta T_{p+1} = T_p$. We shall construct the Φ's in an Euclidean parallelotope P^{2n+1} in the following manner. Let A_{pi} denote the vertices of Φ_p. We construct $\Phi_o = \Phi$ in P^{2n+1} as in (AT,VIII,3.4). Grant that we already have Φ_p. The points $\theta A_{p+1,i}$ are definite vertices of Φ_p', and we construct Φ_{p+1} like Φ save that $A_{p+1,i}$ is to be nearer than a certain η_p to $\theta A_{p+1,i}$. As a consequence by (I,7.14) if $z \in |\Phi_{p+1}|$ then $d(z,\theta z) < \eta_p$. Let α_p be the Lebesgue number of the covering of $|\Phi_p|$ by the stars of the simplexes of Φ_p. We choose $\eta_o = \eta$ arbitrarily and then

(14.1) $$\eta_p < \inf (\tfrac{1}{4}\alpha_p, \tfrac{1}{2}\eta_{p-1}).$$

Consider now the points $y_p = T_p x$ in P^{2n+1}. Since $\theta y_{p+1} = y_p$, we have

(14.2) $$d(y_p, y_{p+1}) < \eta_p < 2^{-p}\eta .$$

Therefore $\{y_p\}$ converges to a point Y of P^{2n+1}. Thus S: x $\longrightarrow$ Y is a transformation of $\mathfrak{R}$ into a subset R of P^{2n+1} and we merely need to prove S topological.

(a) S is one-one. It is only necessary to show that $x \neq x' \Longrightarrow Sx \neq Sx'$. Let y_p', Y' have their obvious meaning and choose p so high that mesh $\mathfrak{U}_p < d(x,x')$. Then no set of $\mathfrak{U}_p$ contains both x,x' and hence by (4.3) no star of Φ_p contains y_p and y_p'. Therefore

(14.3) $$d(y_p, y_p') \geq \alpha_p.$$

On the other hand (15.2) implies

(14.4) $$d(y_p, Y) < \inf\left(\frac{1}{2}\alpha_p, \frac{\eta}{2^{p-1}}\right)$$

and similarly for y'_p, Y'. Therefore $Y = Y' \Longrightarrow d(y_p, y'_p) < \alpha_p$, which contradicts (14.3). Therefore S is one-one.

(b) S is bicontinuous. Since $\mathfrak{R}$ is a compactum we merely need to show that S is continuous. Let $\{x^h\} \rightarrow x$ and set $T_p x^h = y_p^h$, $Tx^h = Y^h$. In view of (I,7.14) we may choose p so high that T_p is an ϵ transformation. Then since T_p is continuous we may choose h such that $d(y_p^h, y_p) < \epsilon$. As a consequence $d(Y^h, Y) < 3\epsilon$, hence $\{Y^h\} \longrightarrow Y$, S is continuous and therefore topological.

The existence of S proves the imbedding theorem.

15. In the preceding argument the dimension plays no role and P^{2n+1} might equally be replaced by any P^q, $q \geq 2n+1$ or by the Hilbert parallelotope P^ω. Suppose then that S_0 is a given mapping of $\mathfrak{R}$ into P^q or P^ω. Choose for each A_{pi} one point B_{pi} on the corresponding set U_{pi} of $\mathfrak{U}_p$ on $\mathfrak{R}$ and let $S_0 B_{pi} = C_{pi}$. We may then carry out the construction so that $d(S\,B_{pi}, C_{pi}) < \epsilon$ throughout. Since $\{B_{pi}\}$ is dense in $\mathfrak{R}$ and the latter is a compactum, we have:

> (15.1) Let S_0 be a mapping of a compactum $\mathfrak{R}$ into P^q, $q \geq 2n+1$, or P^ω. Then there is a topological mapping S of $\mathfrak{R}$ into a subset R of P^q or P^ω as the case may be, such that $d(Sx, S_0x) < \epsilon$ for every $x \in \mathfrak{R}$.

16. The construction of (14) will now be utilized to obtain a deformation very useful later (IV,18). We first modify the construction in that $\mathfrak{R}$ is now mapped into P^ω instead of P^{2n+1}. Then if P^ω is referred to $x_1, x_2, \ldots$, we choose the coordinate x_1 of all the A_{pi} equal to a $\zeta_p > 0$ where $\{\zeta_p\} \longrightarrow 0$ monotonely and rapidly enough not to disturb the inequalities needed

in (14). That this may always be done is seen at a glance. In addition the vertices A_{pi} are to be chosen one at a time such that the matrix of the coordinates of the first q (and hence of any q) is of rank q. It is a consequence of the construction that $R \subset P_1^{\omega}$ where $P_1^{\omega} = \{x \mid x_1 = 0\}$.

We notice also that the covering $\mathfrak{B}$ at the beginning of (14) may be replaced by an irreducible refinement of order $\leqq n$, and hence all the $\mathfrak{U}^{(p)}$ may be chosen irreducible (AT,VII,3.4). As a consequence there are points on U_i^p which are on no other sets of $\mathfrak{U}^{(p)}$ (AT,VII,3.4) and so by (4.1) $T_p\mathfrak{R}$ will contain every vertex A_{pi} of Φ_p. Since $\theta T_{p+1} = T_p$, A_{pi} is the projection of an $A_{p+1,j}$ and hence it is a member of a sequence $\{A_{pi_p}\}$ tending to a point of R. It follows that sup $\{d(A_{pi},R)\} \longrightarrow 0$, and so:

(16.1) $\bigcup \Phi_p$ is an Euclidean complex in $P^{\omega} - P_1^{\omega}$ and regular relative to R.

Consider now the prism $l \times \Phi_{p+1}$, where l is the segment $0 \leqq u \leqq 1$. Having ordered $\{A_{p+1,i}\}$ we proceed as in (AT,VIII,23.3) and if $B_i = 0 \times A_{p+1,i}$, $B_i' = 1 \times A_{p+1,i}$ then we obtain for each $\sigma^r = A_{p+1,i_0} \cdots A_{p+1,i_r} \in \Phi_{p+1}$ with $i_0 < \ldots < i_r$, a set of simplexes $\zeta = B_{i_0} \ldots B_{i_q} B'_{i_q} \ldots B'_{i_s}$ which together with all their faces make up a simplicial decomposition L_{p+1} of $l \times \Phi_{p+1}$. We now define a mapping $\tau_p : |L_{p+1}| \longrightarrow P^{\omega}$ by: $\tau_p B_i = A_{p+1,i}$; $\tau_p B_i' = A_{pi}' = \theta A_{p+1,i}$; τ_p is barycentric. There results a finite Euclidean complex M_{p+1} over which $\theta_p = \theta \mid \Phi_{p+1}$ is now a deformation $|\Phi_{p+1}| \longrightarrow |\Phi_p'|$. If $y \in \sigma^r$ its path $\lambda_p(y) = \tau_p(l \times y)$ is the union of a set of at most $q + 1$ segments, one on each of the simplexes $\tau_p\zeta$. That the number need not be $q + 1$ is due to the fact that some of the

$\tau_p \zeta$ may coincide. However since $q \leqq n$ and θ_p is an η_p mapping we have diam $\lambda_p(y) \leqq (n+1)(\eta_p + \text{mesh } \Phi_p + \text{mesh } \Phi_{p+1}) = \epsilon_p \longrightarrow 0$, and τ_p is an ϵ_p deformation.

Consider now $M = \bigcup M_p$. For $x \in |M_p|$ we have $\zeta_{p+1} \leqq x_1 \leqq \zeta_p$, the equalities holding respectively on Φ_{p+1} and Φ_p. Therefore $M_p \cap M_q = \emptyset$ for $|p-q| > 1$ and $M_p \cap M_{p+1}$ is a subcomplex of Φ'_{p+1}. It follows from (16.1) that M is a polyhedral complex regular relative to R. A simplicial subdivision of the type of (I, 2.2) may now be made replacing M by an Euclidean complex regular relative to R without introducing new vertices and in particular without modifying the Φ's. For simplicity we continue to denote this subdivision by M, and by M_p the related subdivision of the initial M_p. We will set $\tilde{M}_p = \bigcup \{M_q \mid q \geqq p\}$.

We will now define a mapping $S\colon\ 1 \times \mathcal{R} \longrightarrow \mathcal{R} \times M$ as follows:

$$S(0 \times x) = x;\ S(u \times x) = \tau_p[(2^{p+1}u-1) \times T_{p+1}x],$$
$$2^{-p-1} \leqq u \leqq 2^{-p}.$$

It is clear that $S(u \times x) \in |\tilde{M}_p|$ for $u \leqq 2^{-p}$. Therefore T_p is a deformation $R \to \Phi_p$ over $|\tilde{M}_p|$. Since $\epsilon_p \longrightarrow 0$ and $T_p x \longrightarrow x$, T_p is a ξ_p deformation where $\xi_p \longrightarrow 0$.

To sum up then we have:

(16.2) A compactum $\mathcal{R}$ whose dimension $\leqq n$ may be imbedded topologically in a compactum $\mathfrak{S} \subset P^{\omega}$ such that: (a) $\mathfrak{S} - \mathcal{R} = |M|$, where M is a Euclidean complex regular relative to $\mathcal{R}$; (b) whatever ϵ there is an ϵ deformation $\mathcal{R} \longrightarrow \Phi$ over $\mathfrak{S} \cap M_1$, where M_1 and Φ are subcomplexes of M and Φ is finite and at most n dimensional.

§5. RETRACTION

17. The important theory of retraction has been created in the last decade by Borsuk. The preliminary definitions were already given in (AT,I,47), and others will now be added. We confine the treatment to separable metric spaces, as little is known on these questions for more general spaces.

(17.1) DEFINITIONS. We recall that if B is a topological space and $A \subset B$, then a mapping $t: B \longrightarrow A$ such that $T|A = 1$, is known as a retraction of B onto A; if a retraction $B \longrightarrow A$ exists then A is known as a retract of B. The separable metric space A is said to be an absolute retract (abridged into AR) whenever a topological image of A as a closed subset A_1 of any other separable metric space B is necessarily a retract of B. Whenever under the same conditions A_1 is necessarily a neighborhood retract of B then A is said to be an absolute neighborhood retract (abridged into ANR).

(17.2) DEFINITION. Kuratowski [c,d] has generalized the definitions of Borsuk as follows: in (17.1) assume that the retractions only take place when $\dim (B-A_1) \leq n$. The new types obtained are known as absolute n-retracts and absolute neighborhood n-retracts (abridged respectively into AR^n and ANR^n).

(17.3) Remark. Borsuk's original definition for AR and ANR referred only to compacta and may be stated as follows: The compactum A is an AR [an ANR] whenever a topological image A_1 of A in any separable metric space B is necessarily a retract [neighborhood retract] of B. A similar modification holds for n-retraction. Since A_1 is a compactum it is closed in B, and this is the reason why no further condition needs to be placed upon A_1.

(17.4) NOTATIONS. Throughout the remainder of the chapter we write:

P^ω = the Hilbert paralellotope $\mathsf{P}\, l_n$ where l_n is the segment $0 \leqq x_n \leqq \frac{1}{n}$. Wherever a metric occurs it is assumed to be defined by the Euclidean distance (AT,I,44.7).

$P_1^\omega = l_2 \times l_3 \times \ldots$; it is likewise a Hilbert paralellotope;

$P^n = l_1 \times \ldots \times l_n$;

π_n = the projection $P^\omega \to P^n$. That is to say if $x = (x_1, x_2, \ldots)$ then $\pi_n x = (x_1, \ldots, x_n, 0, \ldots)$.

A convenient and often useful preliminary property is:

(17.5) If B, C are separable metric spaces, B is a closed subset of C, and t is a mapping $B \to P_1$, then t has an extension $t_1 \colon C \to P^\omega$ such that $t_1(C-B) \subset P^\omega - P_1^\omega$.

By (AT,I,34.2), t has an extension $t_0 \colon C \to P_1^\omega$, so that $t_0 | B = t$. Modify t_0 to t_1 as follows: for $y \in B$ take $t_1 y = t_0 y$; for $y \in C-B$ and $t_0 y = (0, x_2, \ldots)$ take $t_1 y = (x_1, x_2, \ldots)$, where $x_1 = \dfrac{d(y,B)}{1+d(y,B)}$. Since B is closed in C, necessarily $t_1(C-B) \subset P^\omega - P_1^\omega$, and since $t_1 | B = t_0 | B = t$, (17.5) is proved.

18. We pass to the consideration of Borsuk's theorems.

(18.1) THEOREM. A n.a.s.c. for the separable metric space A to be an absolute retract [absolute neighborhood retract] is the following: if B, C are separable metric, B is a closed subset of C, and there exists a mapping $t: B \longrightarrow A$, then t may be extended to C [to a neighborhood of B in C].

Consider first the AR case. The sufficiency proof is very simple. Let A be a closed subset of the separable metric space D. Under the hypothesis the identical mapping $A \longrightarrow A$ may be extended to D. It yields a mapping $\theta: D \longrightarrow A$ such that $\theta | A = 1$. In other words A is a retract of D, and hence it is an AR.

Passing to necessity let A be topologically imbedded in P_1^ω. By (17.5) t considered as a mapping $B \longrightarrow P_1^\omega$ may be extended to a mapping $\tau_1: C \longrightarrow P^\omega$ such that $\tau_1(C-B) \subset P^\omega - P_1^\omega$. Since P_1^ω is compact and $A \subset P_1^\omega$ necessarily $\bar{A} \subset P_1^\omega$ Therefore $\tau_1 C \subset P^\omega - (\bar{A}-A) = E$ and so $\bar{A} \cap (A \cup \tau_1 C) = A$. Thus A is closed in $A \cup \tau_1 C$ and so it is a retract of $A \cup \tau_1 C$. If τ_2 is the retraction, $t_1 = \tau_2 \tau_1$ is a mapping $C \longrightarrow A$ and $t_1 | B = \tau_1 | B = t$. This disposes of the AR case.

Passing to the ANR case the sufficiency proof is the same save that θ is merely an extension to a neighborhood of A in D. In the necessity proof the argument is essentially the same also save that τ_2 will now be an extension to a neighborhood N of A in E. Since τ_1 is continuous there will be a neighborhood N_1 of B in C such that $\tau_1 N_1 \subset N$, and this time $t_1 = \tau_2 \tau_1 | N_1$ is an extension of t to N_1 mapping N_1 into A. This proves necessity and hence also (18.1).

(18.2) THEOREM. A n.a.s.c. for a compactum to be an absolute retract [absolute neighborhood retract] is that it possess a topological image in P^ω which is a retract [neighborhood retract] of P^ω.

(18.3) Remark. It is a consequence of the theorem itself that whenever its condition is fulfilled for a single topological imbedding of A into P^ω then it is fulfilled for all others.

Proof of (18.2). Necessity is obvious. Regarding sufficiency suppose that t is a topological imbedding of A into P^ω such that $tA = A_1$ is a retract of P^ω. Let A be a subset of the separable metric space B. Since A is compact it is closed in B. Therefore (AT, I,34.2) the mapping $t\colon A \longrightarrow A_1 \subset P^\omega$ may be extended to a mapping $t_1\colon B \longrightarrow P^\omega$. If θ is the retraction $P^\omega \longrightarrow A_1$ then $T^{-1}\theta t_1$ is a retraction $B \longrightarrow A$. This proves (18.3) for retraction.

Suppose now that A_1 is merely a neighborhood retract of P^ω. There exists then a neighborhood N of A_1 in P^ω with a retraction $\theta : N \longrightarrow A_1$. Then $t_1^{-1}N = N_1$ is a neighborhood of A in B retracted by $t^{-1}\theta t_1$ into A. This disposes of the case of neighborhood retraction.

APPLICATIONS (18.4). Every finite polyhedron $|K|$ is an absolute neighborhood retract.

By (AT,VIII,7.5) we may choose $|K| \subset P^n$ and such that there is a retraction ρ of a certain neighborhood N_1 of $|K|$ in P^n, into $|K|$. Then $\rho\pi_n$ retracts the neighborhood $N = \pi_n^{-1} N_1$ of $|K|$ in P^ω, into $|K|$, and so $|K|$ is an ANR.

(18.5) A closed and convex subset A of P^ω is an absolute retract.

If $x \in P^{\omega}$ there is a point y such that $d(x,y) = d(x,A)$, and y is unique. For if $y' \neq y$ had the same property the segment $\overline{yy'}$ would contain a point $z \in A$ nearer than $d(x,A)$ to x. If ρ is the mapping $x \longrightarrow y$, ρ is containuous and retracts $P^{\omega} \longrightarrow A$, hence A is an AR.

19. (19.1) THEOREM. Let A be a closed subset of the Hilbert paralellotope P^{ω}. There may be constructed in $P^{\omega} - A$ on Euclidean complex K which is regular relative to A and is such that $A \cup |K|$ is a retract of P^{ω} (and hence an absolute retract). Furthermore the retraction $T: P^{\omega} \longrightarrow A \cup |K|$ may be chosen such that $P^{\omega} - A$ is deformed over itself onto $|K|$. (LEFSCHETZ [d])

We set $A_n = \pi_n A$. Since π_n is continuous and A is compact so is A_n. We shall introduce for each n certain Euclidean complexes Q_n, R_n, S_n, K_n. The first three are related as follows;

(19.2) $S_1 = l_1$, and for $n > 1$: S_n = a subdivision of the prism $l_n \times R_{n-1}$;

(19.3.) R_n = the set of closures of the simplexes of S_n meeting A_n;

(19.4) $Q_n = \mathrm{Cl}(S_n - R_n)$.

The construction is consistent so far and moreover we may impose

(19.5) mesh $S_n < \frac{1}{n}$ and is so small that no simplex has vertices in both Q_{n-1} and R_n.

We set $Q_n \cap R_n = F_n$ and so owing to (19.5):

(19.6) $F_{n-1} \cap F_n = \emptyset$.

The construction of K_n requires more attention. At all events we choose $K_1 = \emptyset$, and so we may assume $n > 1$. Notice at the outset that we may keep on subdividing our complexes, provided that no element is affected more

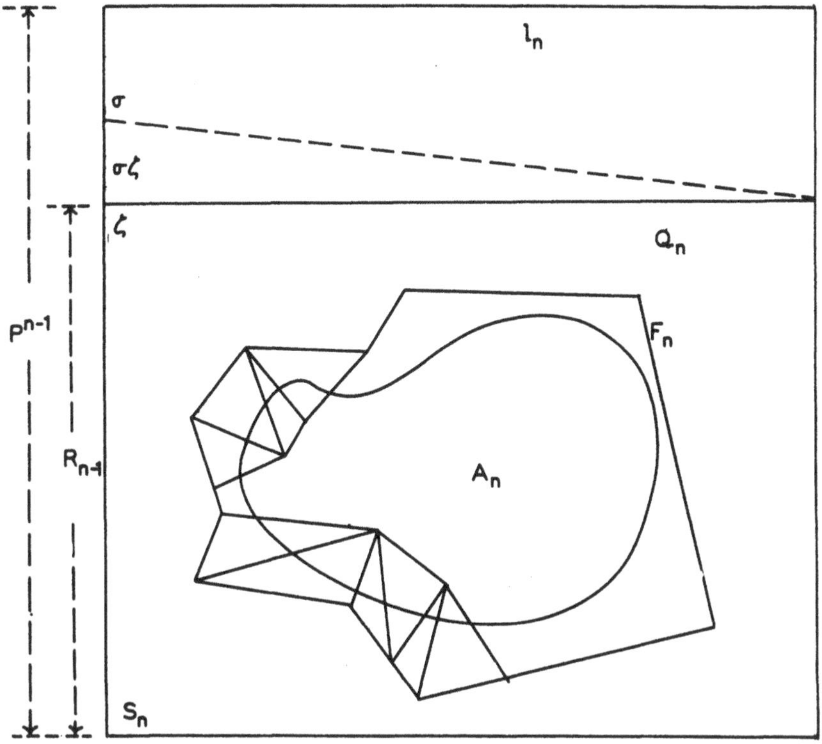

than a finite number of times by the operation. Therefore we may assume at the start R_{n-1} normal in S_{n-1}. Set now $L_{n-1} = \mathrm{St}\, R_{n-1} - R_{n-1} \subset Q_{n-1}$. Every simplex of L_{n-1} is of the form $\sigma\zeta$ where $\zeta \in R_{n-1}$ and $\sigma \in S_{n-1} - R_{n-1}$. Consider now the prism $l_n \times \sigma\zeta$. If $l_n = ab$, let $\sigma\zeta$ be identified with $a \times \sigma\zeta$. Then the simplex $\sigma(b \times \zeta)$ causes a partition of the prism into two polyhedra, one of which denoted by $[\sigma\zeta]$, has the faces $\sigma\zeta$ and $l_n \times \zeta$. We define $\tilde{K}_n$ = the set of all the $[\sigma\zeta]$ together with all the simplexes of Q_{n-1} which are not faces of the $[\sigma\zeta]$. We now subdivide $\tilde{K}_n$ as follows: each $l_n \times \zeta$ is replaced by the set of the elements of S_n covering it; an analogue of barycentric subdivision is applied to the $[\sigma\zeta]$ with new vertices on all faces not in Q_{n-1} The resulting Euclidean complex is K_n. We notice the following properties:

(19.7) $|K_n|$ is a retract of $l_n \times Q_{n-1}$ under a transformation τ_n such that $\pi_{n-1}\tau_n x = \pi_{n-1}x$.

Corresponding to $x \in l_n \times Q_{n-1}$ define $\lambda(x)$ as follows: $\lambda(x) = x$ when $x = \pi_{n-1}x$ or x is not on any $\sigma\zeta$; $\lambda(x) =$ the segment $[\pi_{n-1}x, x]$ otherwise. The set $\lambda(x) \cap K_n$ is a point $x' = \pi_{n-1}(x)$, or else a segment with $x' \neq \pi_{n-1}x$ as an end-point. In all cases we choose $\tau_n x = x'$ and τ_n behaves manifestly in accordance with (19.7).

From the construction follows also:

(19.8) $|K_n| \subset S_n - A_n$;

(19.9) $|K_n| \cap |K_{n+1}| = |l_{n+1} \times F_n|$;

(19.10) $|K_n| \cap |K_{n+2}| = \emptyset$.

(19.11) mesh $K_n \longrightarrow 0$ (since mesh $K_n \leq$ mesh S_n and mesh $S_n \longrightarrow 0$).

20. We now introduce $K = \bigcup K_n$ and show that it satisfies the conditions of the theorem. By (19.10) $|K_n| \cap A_n = |K_n| \cap A = \emptyset$, and so $|K| \subset P^\omega - A$. Since the K_n are finite by (19.10) K is locally finite. Since $S_n \rightarrow A$ from (19.8,19.10) follows then that K is regular relatively to A.

We shall now define the retraction $T: P^\omega \longrightarrow A \cup |K|$ Since $|R_n| \longrightarrow A$, if $x \in P^\omega - A$, there is a first n such that $\pi_n x \notin |R_n|$. Then $\pi_{n-1}x \in |R_n|$ and so $\pi_n x \in |K_n|$, and we chose $Tx = \tau_n \pi_n x$. Notice that this makes $Tx = \pi_{n-1}x$ when $\pi_n x \in |l_n \times F_{n-1}|$ and also $Tx = \pi_n x$ when $x \in |l_{n+1} \times F_n|$. Therefore T is single-valued on

$P^\omega - A$. Since τ_n, π_n are continuous, T is continuous on the closed sets $T^{-1}|K_n|$, and since by (19.10) only consecutive sets $T^{-1}|K_n|$ may intersect, T is a mapping $P^\omega - A \longrightarrow |K|$. In point of fact if $x \in |K_n|$ then $Tx = \pi_n \tau_n x = \pi_n x = x$. Therefore T is a retraction $P^\omega - A \longrightarrow |K|$. Since $Tx \in |K_n|$ implies that the segment joining x to Tx is projected by π_n into the set $\lambda(\pi_n x)$ which does not meet A_n, the segment itself does not meet A. Therefore T is a deformation-retraction $P^\omega - A \longrightarrow |K|$.

To conclude the proof it will be convenient to have:

(20.1) If $Tx \in |K_h|$ then $d(x, Tx) < \rho(h) + \frac{1}{h}$, where $\rho^2(h) = \sum_n (h+n)^{-2}$.

This property is an immediate consequence of the inequalities: $d(x, T_h x) \leqq d(x, \pi_h x) + d(\pi_h x, \tau_h \pi_h x)$; $d(x, \pi_h x) \leqq \rho(h)$; $d(\pi_h x, \tau_h \pi_h x) \leqq \frac{1}{n}$.

Returning then to our theorem suppose $\{x^r\} \longrightarrow A$ in $P^\omega - A$, and let $Tx^r \in |K_{n_r}|$. From our construction follows then that $d(x^r, A) > \inf \{\operatorname{diam} \sigma | \sigma \in S_{n_r+1}\}$ and therefore $n_r \longrightarrow \infty$. Hence by (20.1): $d(x^r, Tx^r) \longrightarrow 0$. Consequently if we set $T = 1$ on A, T becomes a retraction $P^\omega \longrightarrow A \cup |K|$. Notice also that since $d(x^r, Tx^r) \longrightarrow 0$ and the deformation $P^\omega - A \longrightarrow |K|$ is along the segment $\overline{x^r, Tx^r}$, T is also a deformation $P^\omega \longrightarrow A \cup |K|$. This proves (19.1) in all its parts.

21. A part of (19.1) was generalized by Kuratowski [b] to the

(21.1) THEOREM. Given two separable metric spaces A, B with A closed in B and $\dim(B-A) \leqq n$, there exists a mapping S

of B into a subset C of P^ω such that: (a) $S|A$ is topological; (b) $C-SA = |L|$ where L is an Euclidean complex of dimension $\leq n$ and regular relatively to SA.

We may assume at the outset $B \subset P^\omega$. Since A is closed in B we have $\bar{A} \cap (B-A) = \emptyset$. Let K, K_n, T of (19,20) refer now to $\bar{A}$. Since $K \subset P^\omega - \bar{A}$, likewise $K \cap A = \emptyset$. Let $\{a_i\}$ be the vertices of K. Then $\mathfrak{U} = \{U_i\}$, $U_i = B \cap T^{-1}|\text{St } a_i|$, is a countable locally finite covering of B-A. Moreover $\mathfrak{U}$ is regular relatively to A. For the vertex a_i may belong to one K_h, or to two necessarily consecutive say K_h, K_{h+1}. It follows by (20.1) that if $y, y' \in U_i$ then $d(y, Ty)$, $d(y', Ty') < \rho(h) + \frac{1}{h}$. Since $h \to \infty$ with i, we have: $\text{diam } U_i < 2\rho(h) + \frac{2}{h} + \text{diam } |\text{St } a_i| \to 0$. Since both $d(a_i, \bar{A}_i)$ and $d(a_i, U_i) \to 0$ likewise $d(\bar{A}, U_i) \to 0$, and so $\mathfrak{U}$ has the asserted regularity property.

We shall need the following result due to Menger (Dimensionstheorie p.158). The proof which he gives for finite coverings applies to those of the statement below, and is omitted as involving dimension-technique of rather high order.

(21.2) Let $\mathfrak{U} = \{U_i\}$ be a locally finite countable open covering of a separable metric space $\mathcal{R}$ of dimension $\leq n$. There can be found for each U_i an open set $V_i \subset U_i$ such that $\mathfrak{V} = \{V_i\}$ is a covering of order $\leq n$.

Applying (21.2) to our own $\mathfrak{U}$ there is obtained a $\mathfrak{V} = \{V_i\}$ of order $\leq n$. It is a consequence of (21.4) that $\mathfrak{V}$ has for geometric nerve a subcomplex L of the n-section of K so that $\dim L \leq n$. We have: $|L| \subset |K| \subset P^\omega - \bar{A}$, so if $C = |L| \cup A$, A is closed in C, and since K is regular relatively to A, so is L.

Notice also that since C is in P^{ω} it is totally bounded. Let now S be the canonical mapping $B-A \longrightarrow |L|$. Suppose $y \in |[V_1 \dots V_j]|$ (strict intersection). Since $V_1 \subset U_1, \dots$, we shall have $y \in |[U_1 \dots U_j \dots U_k]|$. Since S, T are canonical relatively to $\mathfrak{B}$, we have: $Sy \in \overline{a_1 \dots a_j}$, $Ty \in \overline{a_1 \dots a_j \dots a_k}$ and therefore $d(Sy,Ty) < \operatorname{diam} |\operatorname{St} a_1|$ and so $\longrightarrow 0$ with increasing i. Now when $y \longrightarrow y^0 \in A$, necessarily $i \longrightarrow \infty$ and also $Ty \longrightarrow y^0$, hence $d(y,Ty) \longrightarrow 0$. Therefore $d(y,Sy) \longrightarrow 0$ also. Consequently as in (20) S may be extended to B by choosing $S = 1$ on A. Thus S maps $B \longrightarrow C$ in accordance with (21.1) which is therefore proved.

(21.3) <u>Remark</u>. As a complement to (21.1) we observe that if A is mapped topologically in any manner in P^{ω}, and if the image is still called for convenience A, then L is a subcomplex of the n-section K^n of the complex K of (17.1) constructed relatively to $\bar{A}$. This describes in a sense the degree of freedom in the construction.

22. We are now ready for the comprehensive

(22.1) THEOREM. A separable metric space A may always be imbedded topologically as a closed set, still called conveniently A, of a subset B of P^{ω} such that: (a) $B-A = |K|$, where K is an Euclidean complex regular relatively to A. (b) B is an absolute retract; (c) if K^n is the n-section of K then $A \cup |K^n| = B_n$ (in which A is also closed) is an absolute n-retract.

We may assume at the outset $A \subset P_1^{\omega}$. Since the latter is compact it contains $\bar{A}$, and we construct this time the K of (19.1) for $\bar{A}$ and P^{ω}. Let $B = A \cup |K|$. Once more B is totally bounded and A is closed in B. All that remains to prove therefore is (b), (c). To prove that B is an absolute retract, let C be closed in D, both separable metric, and let C be mapped by τ_0 into B. By (AT,I,34.2) τ_0 may be extended to a mapping $\tau_1 : D \longrightarrow P^\omega$ to be modified presently. First introduce the function $\phi(x)$, $x \in P^\omega$, defined by $\phi(x_1) = 0$ for $x_1 \geqq \frac{1}{2}$, and $\phi(x_1) = \frac{1}{2} - x_1$ otherwise. It is clear that $\phi(x)$ is continuous. Define now $\tau_2 y = \tau_1 y = \tau_0 y$, $y \in C$; if $y \in D-C$ and $\tau_1 y = x = (x_1, x_2, \ldots)$, then choose $\tau_2 y = (x_1', x_2, \ldots)$ where $x_1' = \frac{d(y,C)}{1+d(y,C)}\phi(x_1)+x_1$. Since clearly $0 \leqq x_1' \leqq 1$, τ_2 is still a mapping $D \longrightarrow P^\omega$ and since C is closed in D, no point of D-C is mapped on P_1^ω, hence none on $\bar{A}$. In other words $\tau_2(D-C) \subset P^\omega - \bar{A}$. Moreover clearly $\tau_2 | C = \tau_0$. Apply now the retraction $\theta: P^\omega \longrightarrow \bar{A} \cup |K|$ of (19.1) which maps: $P^\omega - \bar{A} \longrightarrow |K|$. If $\tau = \theta\tau_2$ it is clear again that $\tau | C = \tau_0$ and as $\tau D \subset A \cup |K| = B$, (b) is proved. Notice incidentally that $\tau(D-C) \subset |K|$.

23. Passing to (c) suppose now C, D separable metric, C closed in D, $\dim(D-C) \leqq n$, and C mapped by τ_0 into B_n. It is to be proved that τ_0 may be extended to D. By (21.1) there is a mapping t of D into a subset D_1 of P^ω such that $t|C$ is topological and that $D_1 - tC = |L|$, where L is an Euclidean complex of dimension $\leqq n$ regular relatively to tC. We have then a mapping $\tau_0 t^{-1} : C_1 \longrightarrow B_n$, and it is clear that to extend τ_0 to D it is sufficient to extend $\tau_0 t^{-1}$ to D_1. In other words we may start with the assumptions: (α) $D \subset P^\omega$; (β) $D-C = |L|$, L an Euclidean complex of dimension $\leqq n$ and regular relatively to C.

We may now define τ as in (20), and we shall have $\tau | C = \tau_0$, $L \subset K$. Thus we have a continuous complex $\mathfrak{L} = (L,\tau) \subset K$, and its properties will be utilized to the full.

Let us set $E = \overline{A \cup \tau_0 C}$. We first cover $|K| - E$ by the same construction as in (AT,VIII,5.1) with a locally finite complex K, here clearly regular relatively to E. The operation consists, as we recall, in taking the barycentric derived $K^{(p)}$ then removing from K the closed simplexes meeting E, and operating in the same way upon the first derived of the complex removed, etc., following finally with a single operation of barycentric derivation. Since τ is continuous $(\tau^{-1}E) \cap D = G$ is closed and so $|L| - G$ may likewise be covered with a locally finite complex $\tilde{L}$ which is regular relatively to G. We have then in $\tilde{\mathfrak{C}} = (\tilde{L}, \tau)$ a continuous complex in K.

(23.1) Each element of one of $\tilde{K}, \tilde{\mathfrak{C}}$ meets at most a finite number of elements of the other.

If $\sigma \in \tilde{K}$, $\bar{\sigma}$ does not meet E, hence $\tau^{-1}\bar{\sigma}$ does not meet G. Since τ is continuous $\tau^{-1}\bar{\sigma}$ is closed in D and so its points are at a positive distance from G. Since $\tilde{L}$ is regular relative to G, $\tau^{-1}\bar{\sigma}$ meets at most a finite number of elements of L. It follows that σ, hence any element of K, meets at most a finite number of elements of $\tilde{\mathfrak{C}}$.

Take now $\zeta \in \tilde{L}$. Since $\bar{\zeta}$ is compact $\tau\bar{\zeta} \subset |\tilde{K}|$ is also compact. Therefore $d(\tau\bar{\zeta}, E) > 0$, and so $\tau\bar{\zeta}$ meets at most a finite number of elements of K. Thus (23.1) is proved.

Since (23.1) holds by (AT,VIII,23.7) there exists a homotopy t_0 of $\tilde{\mathfrak{C}}$ into a subcomplex of the n-section $\tilde{K}^n$ of $\tilde{K}$. We shall consider this subcomplex as a polyhedral complex Π of dimension $\leq n$. Referring to loc. cit. t_0 is such that if x is in σ of K then t_0x is in $\bar{\sigma}$ and so is the path of x in the homotopy. It follows that t_0 may be extended to a homotopy t of $D = G \cup |\tilde{L}|$ into $|\tilde{K}^n| \cup E$ such that $t \mid G = 1$. Therefore τ may be replaced by $t\tau$ which maps $|\tilde{L}|$ into some $|\tilde{L}_1|$, L_1 a subcomplex of K^n. We may thus assume that τ itself maps $|L|$ into $|L_1|$, L_1 a subcomplex of $\tilde{K}^n$.

We are then facing the following situation: there is a mapping $\tau: D \longrightarrow B$ such that $\tau | C = \tau_0$, and that $\tau D \cap (B - |K^n|)$ is a polyhedron Π, whose dimension $\leq n$ and which is regular relatively to E. Take then any $\sigma \in K$ which is its own star and whose dimension $> n$. There may be chosen on σ a point $x \notin |\Pi|$. Since Π is regular relatively to E, we have $d(x, |\Pi|) > 0$. Therefore the radial projection from $x: \bar{\sigma} - x \longrightarrow |\mathfrak{B}\sigma|$ is continuous on $\sigma \cap |\Pi|$ and replaces $|\Pi|$ by a similar polyhedron without points on σ. Repeat the same process for all simplexes such as σ, then remove these from K, and operate again in the same way on the complex left. Since K is locally finite each $\bar{\sigma} \in K - K^n$ undergoes at most a finite number of operations. The total result is obviously a retraction: $|\Pi| \longrightarrow |K^n|$, and hence a retraction $\tau(D-C) \longrightarrow |K^n|$. Since $\tau | C = 1$, the mapping $\tau_0: C \longrightarrow B_n$ has been extended to D, and (22.1 c), hence (22.1) is proved.

(23.2) <u>Remark</u>. We have seen (19.1) that when A is a compactum in P^ω it is always possible to construct K in $P^\omega - A$ itself. Suppose A merely separable metric and in P^ω. We may then replace in the construction just given P^ω by $l \times P^\omega$, where l is a segment, and P_1^ω by P^ω and then K will be in $l \times P^\omega - (\bar{A} - A)$.

24. The theorem just proved will enable us to extend Borsuk's theorems (18.1, 18.2), to separable metric spaces for the last, and with Kuratowski, to n-retraction for both.

(24.1) **THEOREM.** A n.a.s.c. for a separable metric space A to be an absolute n-retract [absolute neighborhood n-retract] is the following: if B, C are separable metric spaces, with B closed in C and $\dim(C-B) \leqq n$, and if there exists a mapping $t: B \longrightarrow A$ then t may be extended to C [to a neighborhood of B in C].

Sufficiency is proved as for (18.1) but necessity is less immediate. As a preliminary step we first prove:

(24.2) Let $A \subset P_1^\omega$ and let $C \subset P^\omega$ and such that $C-B = |L|$ where L is an Euclidean complex of dimension $\leqq n$ regular relative to B. Then t may be extended to a mapping $t_1: C \longrightarrow P^\omega$ such that t_1 is a topological barycentric mapping $|L| \longrightarrow |M|$, where M is an Euclidean complex in $P^\omega - P_1^\omega$ which is regular relative to tB.

Let $\{P_i\}$ be the vertices of L. For each P_i choose a point P_i' on B such that $d(P_i, P_i') < d(P_i, B) + 1/i$. Then in $P^\omega - P_1^\omega$ choose one at a time points Q_i one for each P_i such that: (a) $d(Q_i, tP_i') < 1/i$; (b) the rank of the matrix of the coordinates of $Q_1, \ldots, Q_i$ is i throughout. If $P_j, \ldots, P_k$ are the vertices of a simplex σ of L we construct the simplex ζ in P^ω with the vertices $Q_j, \ldots, Q_k$ and since $P^\omega - P_1^\omega$ is convex we have $\zeta \subset P^\omega - P_1^\omega$. The aggregate $M = \{\zeta\}$ is manifestly an Euclidean complex such that $|M|$ is the topological transform of $|L|$ under a transformation t' which is barycentric topological. Finally if $\{x_p\} \longrightarrow x \in B$ on C, where $x_p \in |L|$, then $\{t'x_p\} \longrightarrow tx$ on $|M| \cup tB$. Since $|M| \cap tB = \emptyset$, t' may be extended to a mapping $t_1: C \longrightarrow |M| \cup tB$ by choosing $t_1|B = t$.

(24.3) Under the same situation as in (24.2) if A is an absolute n-retract then

t may be extended to a mapping $C \longrightarrow A$ (i.e. 24.1 holds).

Since $|M| \subset P^\omega - P_1^\omega$, it does not meet $\bar{A}$ and so $(|M| \cup A) \cap \bar{A} = A$. Therefore A is closed in $|M| \cup A$, and since $\dim M = \dim L \leqq n$, there is a retraction $\theta : |M| \cup A \longrightarrow A$. Hence θt_1 is a mapping $C \longrightarrow A$ extending t, and proving (24.3).

(24.4) We now take the general case of (24.1). By (21.1) there is a mapping $\tau : C \longrightarrow C'$ such that $\tau | B$ is topological and that if $\tau B = B'$ then L', C', B' are related like L, C, B in the case just considered. By (24.3) the mapping $t\,\tau^{-1} : B' \longrightarrow A$ may be extended to a mapping $\tau_1 : C' \longrightarrow A$. Accordingly $\tau_1 \tau$ extends the mapping t and the proof of (24.1) for AR is complete.

(24.5) For ANR the first modification required is in θ which is now merely a retraction of a neighborhood N_1 of A in $|M| \cup A$. Then $N = t_1^{-1} N_1$ is a neighborhood of B in C to which θt_1 extends t. This is the analogue of (24.3) for ANR. Regarding (24.4) τ_1 is now a mapping of a neighborhood N' of B' in C' into A and $N = \tau^{-1} N'$ is a neighborhood of B in C to which $t\,\tau^{-1}$ extends t. This disposes of the ANR case.

25. (25.1) **THEOREM.** A n.a.s.c. for a separable metric space A to be an absolute retract [absolute neighborhood retract] is that it possess a topological image in P^ω, still called conveniently A, such that A is a retract [neighborhood retract] of $E = I \times P^\omega - (\bar{A} - A)$, I a segment. Similarly with retraction replaced by n-retraction.

(25.2) <u>Remark</u>. Same as (18.3): if the condition of the theorem is fulfilled for a single topological imbedding of A into P^ω, it is fulfilled for all others.

Proof of (25.1). Necessity is immediate. We have $E \cap \bar{A} = E \cap A$, hence A is closed in E, and so if it is AR or ANR it must be a retract or neighborhood retract of E.

Suppose now the condition for an AR fulfilled. Referring to (23.1, 23.2) we may construct K in $1 \times P^\omega - \bar{A}$ regular relative to A and such that $A \cup |K|$ is an AR. Suppose now that B is a separable metric space containing a closed subset A_1 which is topologically equivalent to A, and let t be a topological mapping $A_1 \longrightarrow A$. Since $A \cup |K|$ is an AR we may extend t to a mapping $t_1 : B \longrightarrow A \cup |K|$. If θ is the retraction $E \longrightarrow A$ then θt_1 is a mapping $B \longrightarrow A$ which extends t. Therefore A is an AR.

Suppose only the ANR condition fulfilled. Then there is a neighborhood N of A in E with a retraction $\theta: N \longrightarrow A$. Therefore $N' = t_1^{-1}(N \cap t_1 B)$ is a neighborhood of A in B over which θt_1 extends t as a mapping into A. Therefore this time A is an ANR.

The treatment of AR^n and ANR^n is the same with the n-section K^n of K in place of K, and so we omit it.

26. APPLICATIONS (25.1) The n-cell and n-paralellotope are absolute retracts.

For the paralellotope it is a consequence of (18.4). For the n-cell it may be proved as follows. Let P^n, π_n be as in (19.1) and let S^{n-1} be the boundary-sphere of P^n. We shall show that the n-cell $E^n = P^n - S^{n-1}$ is a retract of $P^\omega - S^{n-1}$. It will follow by (25.1) that E^n is an AR.

Given any $x \in P^\omega - P^n$ and $\xi \in S^{n-1}$, we have $d(x, \xi) \geqq d(x, P^n)$. Therefore there exists on ξx a point x' such that $d(x, x') = d(x, P^n)$. Define now a mapping τ: $P^{\omega} - P^{n} \longrightarrow P^{\omega}$ by $\tau x = x'$, $x \notin P^{n}$. Since P^{n} is closed in P^{ω}, if we choose $\tau = 1$ on P^{n}, τ becomes a mapping of P^{ω} into itself. It is easy to verify that $d(\xi, \pi_n \tau x) < d(\xi, \pi_n x)$ for every $x \in P^{\omega} - P^{n}$. Therefore $\pi_n \tau$ maps $P^{\omega} - S^{n-1} = P^{\omega} - (\bar{E}_n - E_n)$ into E^{n} and as $\pi_n \tau | E_n = 1$, E^{n} is an AR.

Chapter IV

LOCAL CONNECTEDNESS AND RELATED TOPICS

The spaces dealt with in the present chapter are all characterized by simple local homotopy or homology properties. Their prototype is the <u>continuous curve</u> (image of a segment under a mapping) whose treatment is readily available in the literature and so will receive little more than a passing mention.

In its present status the theory may as well be confined to compacta, and so until the last section (a mere summary of certain generalizations and of generalized manifolds), all sets are supposed to be separable metric. Furthermore all the properties under discussion are manifestly topological and so no further mention of the fact will be made in the sequel.

§1. LOCALIZATION

1. "Localization" is a generic term applied to various processes for introducing a property in the neighborhood of points of a space, (local property) analogous to some known property of the space as a whole (property in the large). Thus compactness has given rise to local compactness, and from connectedness and certain similar properties we shall arrive at the general type of LC properties (LC abridged for "local connectedness"). The first instance of the properties in question arose out of the search for an intrinsic characterization of

continuous curves. A few words regarding these will serve as a good introduction to our present topic.

(1.1) DEFINITION. A space $\mathfrak{R}$ is said to be locally connected at the point x whenever every neighborhood U of x contains another V such that if $x_1, x_2 \in V$ then there is a connected set $A \subset U$ containing both points. The space $\mathfrak{R}$ is said to be locally connected, whenever it is locally connected at every point.

We state the important

(1.2) THEOREM. A continuous curve is characterized by the following properties: it is a connected and locally connected compactum. (Hahm - Mazurkiewicz).

For a proof (largely due to Sierpinski) see F. Hausdorff: Mengenlehre (2nd Edition) pp. 205-208.

(1.3) DEFINITION. A topological space $\mathfrak{R}$ is said to be arc-wise connected whenever any two points $x, y \in \mathfrak{R}$ may be joined by an arc in $\mathfrak{R}$.

By localization we obtain:

(1.4) DEFINITION. The space $\mathfrak{R}$ is locally arc-wise connected at the point x whenever every neighborhood U of x contains another V such that any two points of V may be joined by a closed arc (closed one-cell) in U. Finally $\mathfrak{R}$ is locally arc-wise connected whenever it is locally arc-wise connected at every point.

(1.4) THEOREM. Every continuous curve is arc-wise connected as well as locally arc-wise connected (R. L. Moore: Trans. Am. Math. Soc. vol. 17 (1916) p. 131).

2. Let a segment $l : 0 \leqq u \leqq 1$ be mapped into a continuous curve λ so that 0 goes into x and 1 into y. We may consider λ as a <u>path</u> from x to y. The existence of λ is equivalent to the assertion: the continuous 0-sphere S^0 consisting of x and y is homotopic to a continuous zero-sphere S^0_1 consisting of a mapping of a zero-sphere into a single point. This statement is also abridged to "S^0 is homotopic to a point". In seeking to generalize this property one may replace the zero-sphere by other spheres or sets. Hence the following definitions:

(2.1) DEFINITION. A set A on the topological space $\mathfrak{R}$ is said to be <u>contractible</u> in $\mathfrak{R}$ whenever the identity $A \longrightarrow A$ is homotopic over $\mathfrak{R}$ to a mapping of A into a point of $\mathfrak{R}$. This is often described more briefly as: "the identity is homotopic over $\mathfrak{R}$ to a constant", or also "A is deformable over $\mathfrak{R}$ into a point". Whenever $\mathfrak{R}$ is contractible in itself we merely say "$\mathfrak{R}$ <u>is contractible</u>".

(2.2) DEFINITION. The space $\mathfrak{R}$ is said to be a p - C <u>space</u>, or to have the p-C <u>property</u>, whenever every continuous p-sphere S^p in $\mathfrak{R}$ is homotopic to a point.

An equivalent definition is: every continuous p-sphere S^p bounds a continuous (p+1)-cell E^{p+1} in $\mathfrak{R}$.

For if $S^p = (\mathfrak{B}\sigma^{p+1}, t)$ is contractible to x in $\mathfrak{R}$ there is a mapping $t_1 . |1 \times \mathfrak{B}\sigma^{p+1}| \longrightarrow \mathfrak{R}$ where 1 is the segment $0 \leqq u \leqq 1$, such that t_1 agrees with t on $|0 \times \mathfrak{B}\sigma^{p+1}|$ and $t_1 | 1 \times \mathfrak{B}\sigma^{p+1}| = x$. Now $|1 \times \mathfrak{B}\sigma^{p+1} - 0 \times \sigma^{p+1}|$ is readily shown to be a $(p+1)$-cell E^{p+1}. It follows that if we identify $\mathfrak{B}\sigma^{p+1}$ with $0 \times \mathfrak{B}\sigma^{p+1}$ then the continuous cell (E^{p+1}, t_1) has the boundary S^p. Conversely to say that S^p bounds a continuous E^{p+1} means that t may be extended to a mapping $t_1 : \bar{\sigma}^{p+1} \longrightarrow \mathfrak{R}$. If s is a deformation of $\bar{\sigma}^{p+1}$ into a point then $t_1 s$ is a homotopic to a constant and so $\mathfrak{R}$ is $p - C$.

(2.3) DEFINITION. The space $\mathfrak{R}$ is said to be a C^p-<u>space</u> or to have <u>property</u> C^p if it is a $q - C$ space for every $q \leqq p$. It is said to be a C^ω <u>space</u> or to have <u>property</u> C^ω if it is a $q - C$ space for every q.

A C^0 space is the same as a $0 - C$ space or again as an arc-wise connected space. For property C^0 implies the existence of a continuous curve joining any two points x, y and by (1.4) they may also be joined by an arc. The converse being obvious our assertion follows.

(2.4) A retract of a contractible space is contractible. Hence a compactum which is an absolute retract is contractible.

Let θ be a retraction $S \longrightarrow R$ where S is contractible. By hypothesis there exists a deformation ω of S into a point x. Then $\theta\omega | R$ is a deformation of R into the point θx. If R is an AR it may be immersed topologically in the Hilbert parallelotope P^ω and since P^ω is contractible so is R.

3. The localization of the concepts just considered leads to the following definitions, in which LC is abridged for "locally connected".

(3.1) DEFINITION. The topological space $\mathfrak{R}$ is said to be $\left\{\begin{array}{l}\underline{\text{locally contractible}}\\ \text{p - LC}\end{array}\right.$ at x whenever every neighborhood U of x contains another V, such that $\left\{\begin{array}{l}\text{V}\\ \text{every continuous p-sphere in V}\end{array}\right.$ is homotopic to a point over U. The space is LC^p at x [LC^ω at x] whenever it is q - LC for every $q \leq p$ [for every q]. Finally $\mathfrak{R}$ is said to be locally contractible, p - LC, LC^p, LC^ω whenever it has the corresponding property for every x.

(3.2) Metric spaces. When $\mathfrak{R}$ is metric equivalent formulations may be given with U, V replaced by spheres $\mathfrak{S}(x,\epsilon)$, $\mathfrak{S}(x, \eta(\epsilon, x))$. Whenever corresponding to any one of the above properties $\eta(\epsilon,x)$ may be chosen independently of x, $\mathfrak{R}$ is said to be uniformly locally contractible, ...

(3.3) Notice that when $\mathfrak{R}$ is compact "local connectedness" is the same as 0 - LC or LC^0.

(3.4) When a compactum is locally contractible, ... it is uniformly locally contractible ... . (Obvious).

As a consequence the following formulation is possible for a compactum:

(3.5) The compactum $\mathfrak{R}$ is $\left\{\begin{array}{l}\underline{\text{locally contractible}}\\ \text{p - LC}\end{array}\right.$ whenever given any $\epsilon > 0$ there is an $\eta > 0$ such that $\left\{\begin{array}{l}\text{every set}\\ \text{every continuous p-sphere}\end{array}\right.$ on $\mathfrak{R}$ of diameter $<\eta$ is ϵ homotopic to a point.

4. (4.1) Local contractibility $\Longrightarrow LC^{\omega} \Longrightarrow LC^{p}$.

(4.2) If a connected compactum $\mathcal{R}$ is locally contractible, LC^p or LC^{ω}, it is a continuous curve (1.2, 4.1).

(4.3) A product of a finite set of LC^p compacta is an LC^p compactum. (Kuratowski)

Let $\mathcal{R} = \mathcal{R}_1 \times \dots \times \mathcal{R}_n$, where $\mathcal{R}_i$ is an LC^p compactum. We must show that $\mathcal{R}$ is q - LC for every $q \leqq p$. Let π_i be the natural projection $\mathcal{R} \longrightarrow \mathcal{R}_i$. If $d_i(x_i, y_i)$ is a metric for $\mathcal{R}$, by (AT, I, 43.8) we may choose for $\mathcal{R}$ the metric

$$d(x,y) = \sum d_i(x_i, y_i).$$

Let $\eta_i(\epsilon, q)$ be the function of the q - LC condition for $\mathcal{R}_i$ and let $\eta(\epsilon, q) = \inf\{\eta_i(\frac{\epsilon}{n}, q)\}$. Consider now an Euclidean simplex σ^{q+1} and let $\zeta = (\mathfrak{B}\sigma^{q+1}, t)$ be an image of $\mathfrak{B}\sigma^{q+1}$ in $\mathcal{R}$ whose diameter $< \eta(\epsilon, q)$. Then $(\mathfrak{B}\sigma^{q+1}, \pi_i t)$ is an image in $\mathcal{R}_i$ whose diameter $< \eta_i(\frac{\epsilon}{n}, q)$, and so $\pi_i t$ may be extended to a mapping $t_i : \bar{\sigma}^{q+1} \longrightarrow \mathcal{R}_i$ such that diam $t_i \bar{\sigma}^{q+1} < \frac{\epsilon}{n}$. Define now a mapping $\tau : \bar{\sigma}^{q+1} \longrightarrow \mathcal{R}$ by the condition that if $z \in \bar{\sigma}^{q+1}$ then $\tau z = t_1 z \times \dots \times t_n z$. We have: $\tau | \mathfrak{B}\sigma^{q+1} = t$, diam $\tau\sigma^{q+1} < \epsilon$. Hence $\mathcal{R}$ is q - LC, proving (4.3).

(4.4) Convex subsets of an Euclidean or Hilbert parallelotope are locally contractible (AT, I, 47).

(4.5) A neighborhood retract of a locally contractible set is locally contractible. Hence a compactum which is an absolute neighborhood retract is locally contractible.

Let S be locally contractible and R a neighborhood retract of S. There exists then a neighborhood N of R in S with a retraction $\theta : N \longrightarrow R$. Let U be a neighborhood of $x \in R$ in R. There exists a neighborhood U' of x in N such that $U' \cap R = U$. Since U' is open in S there exists a second neighborhood V' of x in S with a deformation ω over U' into a point y of U'. Then $V' \cap R = V$ is a neighborhood of x in R deformed by $\theta\omega \mid R$ over $U \cap R$ into the point θy, so R is locally contractible. If R is an ANR it has a topological image R' in P^{ω} which is a neighborhood retract of P^{ω}. Since P^{ω} is locally contractible so is R' and hence also R.

(4.6) Finite polyhedra are locally contractible. (4.4;III,18.4).

(4.7) An n-cell is locally contractible and Euclidean and Hilbert parallelotopes are uniformly locally contractible.

<u>Examples</u> (4.8) The classical example of a space which is not locally connected (= not 0 - LC), is the set of (AT,VIII,25.1) based on the curve $y = \sin\frac{1}{x}$. This space is not a continuous curve.

(4.9) (Wilder). $\mathfrak{R}$ is the spherical region $x^2 + y^2 + z^2 < 1$ from which a "spine" say $y = z = 0$, $0 \leq x < 1$ has been removed. Then $\mathfrak{R}$ is uniformly 0 - LC, also 1 - LC but not uniformly 1 - LC, since a very "small" loop around the spine cannot be shrunk to a point without a "large" displacement.

§2. PARTIAL REALIZATION OF COMPLEXES. APPLICATION TO LOCAL CONNECTEDNESS

5. Let K be a complex. A closed subcomplex L

is said to be dense in K whenever each element of K has a face in L. When K is simplicial L is dense if it contains all the vertices of K.

Suppose now that K, L are finite Euclidean complexes. Then a continuous complex $\mathfrak{K}_1 = (L, t)$ on the metric space $\mathfrak{R}$, is known as a partial realization of K in $\mathfrak{R}$. If $K = \{\sigma\}$, then $\sup \operatorname{diam} t(|L \cap \mathrm{Cl}\,\sigma|)$ is by definition the mesh of the partial realization. Notice that this mesh is quite different from the mesh of the continuous complex $\mathfrak{L} = (L, t|L)$. Indeed this is the only distinction between $\mathfrak{K}_1$ and $\mathfrak{L}$. As an example if K is a plane triangle and L the set of its vertices then mesh $\mathfrak{L} = 0$ while mesh $\mathfrak{K}_1$ as a partial realization is not zero.

Is it possible to supply the missing faces, thus replacing $\mathfrak{K}_1$ by a full image $\mathfrak{K}$ of K? That is to say does t admit an extension to $|K|$? Furthermore given $\epsilon > 0$ can we find an $\eta > 0$ such that when mesh $\mathfrak{K}_1 < \eta$ then there may be obtained an image $\mathfrak{K}$ of mesh $< \epsilon$? When $K = \mathrm{Cl}\,\sigma^{p+1}$, $L = \mathfrak{B}\,\sigma^{p+1}$ we recognize that the questions proposed correspond to the p - C and p - LC properties. The two theorems to be proved presently do in fact establish the closest possible connection between the extension and these two properties.

6. (6.1) THEOREM. A n.a.s.c. in order that every partial realization of a finite Euclidean complex of dimension $\leq p + 1$ may be extended to a full realization is that the space $\mathfrak{R}$ be C^p. A n.a.s.c. in order that the extension may be possible for all p is that the space be C^ω.

(6.2) THEOREM. When $\mathfrak{R}$ is a compactum, a n.a.s.c. in order that given any $\epsilon > 0$ there exist an $\eta > 0$ such that every partial realization of mesh $< \eta$ of a finite Euclidean complex of dimension $\leq p + 1$ may be extended to a full realization of mesh $< \epsilon$, is that the space be LC^p. (Lefschetz [d]).

The proofs of both theorems are given below (7.8).

The function $\eta(\epsilon,p)$ which occurs in the second theorem is said to be an <u>extension function</u> for the space $\mathfrak{R}$. One will notice that while p may be infinite in (6.1), it is to be kept finite in (6.2). In fact if $p = \infty$ is admissible then η may be chosen a function of ϵ alone for all p, and this may not be possible as simple examples show.

(6.3) DEFINITION. We say that the compactum $\mathfrak{R}$ is an LC* <u>space</u>, or that it has the LC* <u>property</u>, whenever given any $\epsilon > 0$ there exists an $\eta(\epsilon)$ (an extension-function of $\mathfrak{R}$) such that every partial realization of a finite Euclidean complex K, whose mesh $< \eta$ may be extended to a full realization whose mesh $< \epsilon$.

(6.4) DEFINITION. A compactum $\mathfrak{R}$ which is both $\left\{\begin{matrix} LC^p \text{ and } C^p \\ LC^* \text{ and } C^\omega \end{matrix}\right.$ is said to be an $\left\{\begin{matrix} LC^{pp} \\ LC^{**} \end{matrix}\right.$ <u>space</u>, or to have the $\left\{\begin{matrix} LC^{pp} \\ LC^{**} \end{matrix}\right.$ property.

(The reader will recall that: (a) C^p means that every continuous q-sphere, $q \leq p$, is contractible; (b) C^ω means that every continuous sphere is contractible.)

An LC^{pp} space is characterized as follows: Every partial realization of an at most p + 1-dimensional finite Euclidean complex may be extended to a full realization, and this may also be done with the ϵ, η restrictions. For the LC** it is the same without dimensional restrictions. It should be noted that LC* implies LC^{ω} but not C^{ω}. For the first only asserts the possibility of extension when the mesh η is small enough, while the second that it may always be carried out without regard to the mesh.

(6.5) As we shall see in the sequel the significant categories are LC^{p}, LC*, LC^{pp} and LC**. This does not signify however that they are the only known categories. From this point of view the following example, communicated to the author by Borsuk, is instructive: Consider in a Hilbert parallelotope a sequence of mutually exclusive spheres $\{S^n\}$, where $\dim S^n = n$, $\operatorname{diam} S^n \longrightarrow 0$ and $S^n \longrightarrow$ a point. The union of the spheres is LC^{ω}. It is readily shown that the function $\xi(\epsilon,p)$ of the p - LC condition $\longrightarrow 0$. Hence the set is not LC*. We shall find that every Euclidean complex is LC*. However an LC** must be contractible. Hence a circumference is LC* but not LC**. Thus while the categories LC^{ω}, LC*, LC** overlap, they are nevertheless distinct.

We may of course define a space which is both LC^{ω} and C^{ω} as $LC^{\omega\omega}$, but we have no information regarding the position of this class relative to the others.

To orient the reader we state immediately our most significant result which will be proved later as part of (12.2):

(6.6) THEOREM. LC* compacta are the same as absolute neighborhood retract compacta. LC** compacta are the same as absolute retract compacta. (Lefschetz [d]).

7. <u>Proof of (6.1)</u>. (a) Let $\mathfrak{R}$ possess the extension property of the theorem. A continuous q-sphere

S^q in $\mathfrak{R}$, $q \leqq p$, is a full realization of a $\mathfrak{B}\sigma^{q+1}$ and hence a partial realization of a $\mathrm{Cl}\,\sigma^{q+1}$. This partial realization may be extended to a full realization of $\mathrm{Cl}\,\sigma^{q+1}$, which means that there exists a continuous cell E^{q+1} bounded, by S^q. Hence $\mathfrak{R}$ is q - C for every $q \leqq p$, that is to say it is C^p, and so the extension property implies that $\mathfrak{R}$ is C^p.

(b) Conversely let $\mathfrak{R}$ be C^p and let K, $\mathfrak{K}_1$ be as before with dim $K \leqq p + 1$. If $\{\sigma_i^1\}$ are the non-realized one-simplexes of K, the spheres $\mathfrak{B}\,\sigma_i^1$ are realized and hence, since $\mathfrak{R}$ is C^p, the images of the σ_i^1 may be inserted. We now reason similarly with the missing simplexes of the successive dimensions 2,3, ... , p+1, and obtain thus a full image $\mathfrak{K}$ of K, so that the C^p property implies the extension property. Hence both are equivalent.

The preceding argument applies even when $p = \infty$, the extension being then possible for every q. Therefore the theorem is completely proved.

8. Proof of (6.2). (a) Let $\mathfrak{R}$ have the extension property with $\eta(\epsilon,p)$ as the extension function. Under the circumstances in (7a) if diam $S^q < \eta(\epsilon,p)$, the extension yields an E^{q+1} whose diameter $< \epsilon$. Therefore $\mathfrak{R}$ is q - LC, for every $q \leqq p$, and hence it is LC^p. Thus the extension property implies that $\mathfrak{R}$ is LC^p.

(b) Passing to the converse assuming $\mathfrak{R}$ to be LC^p, we must produce an extension function $\eta(\epsilon,p)$. Let this time $\xi(\epsilon,q)$ be the function of the q - LC property where $q \leqq p$. Let $\zeta_0, \ldots, \zeta_{p+1}$ be such that

$$(8.1)_q \quad 0 < 2\,\zeta_q < \xi(\zeta_{q+1}, q),\ q \leqq p;$$

$$(8.1)_{p+1} \quad 0 < \zeta_{p+1} < \epsilon.$$

Since $\xi(\epsilon,r) < \epsilon$ we have:

$$(8.2) \quad 0 < \zeta_0 < \ldots < \zeta_{p+1} < \epsilon.$$

Choosing now any positive $\eta(\epsilon,p) < \zeta_0$, we will prove that $\mathfrak{R}$ is LC^p with $\eta(\epsilon,p)$ as extension function. Let in fact K, of dimension $\leq p+1$ have a partial realization $\mathfrak{K}_1$ of mesh $< \eta(\epsilon,p)$. Owing to (8.2) it is sufficient to show that we may insert the missing cells, dimension by dimension, in such manner that those of dimension q are of diameter $< \zeta_q$. The zero-cells are all present, so we may assume that we have already inserted in the requisite way all those of dimension $\leq q$, and we merely have to prove that the missing (q+1)-cells may be inserted to diameters $< \zeta_{q+1}$.

The cells already present make up a partial realization $\mathfrak{K}_2 = (L,t)$ of. K, where L is closed and dense in K. If $\sigma^{q+1} \subset K - L$ and so is not yet mapped, then $\mathfrak{B}\sigma^{q+1} \subset L$ and t is defined on $|\mathfrak{B}\sigma^{q+1}|$. If $\sigma^r \prec \sigma^{q+1}$ then by hypothesis diam $t\sigma^r < \zeta_q$. Since any two faces of σ^{q+1} have a vertex in common we have diam $t|\mathfrak{B}\sigma^{q+1}| < 2\zeta_q$ and hence mesh $\mathfrak{K}_2$ (as a partial realization) is $< 2\zeta_q$. Owing to $(8.1)_q$ we may then insert the missing image of σ^{q+1} to diameter $< \zeta_{q+1}$. This completes the proof of (6.2).

(8.3) In the preceding treatment continuous complexes may be replaced by mappings of polyhedra and their partial realizations. The only modification required is replacing $2\zeta_q$ in $(8.1)_q$ by $3\zeta_q$. In point of fact we shall only need <u>continuous prisms</u> (mappings of prisms), and partial realizations of prisms (defined as in 5). We state explicitly:

(8.4) Both (6.1) and (6.2) hold with Euclidean complexes replaced by prisms.

(8.5) If $\eta(\epsilon,p)$ is an extension function for Euclidean complexes then $\eta(\frac{1}{3}\epsilon,p)$ is one for prisms. Therefore if a space $\mathfrak{R}$ is LC*, LC^{pp}, LC**, in the earlier sense, it remains so when

Euclidean complexes are replaced everywhere by prisms.

The arguments of (7) and (8a) may be repeated with minor modifications. Let now Π be a prism with a partial realization Ω whose mesh (as a partial realization) is $< \eta(\frac{1}{3}\epsilon, p)$. Let ω be any unmapped cell of Π. Take a simplicial subdivision K of Π such as in (AT,VIII, 23.3) without any new vertices. Then the derived Ω' constitutes a partial realization $\mathfrak{K}_1$ of K whose mesh $< \eta(\frac{1}{3}\epsilon, p)$ and so it may be completed to a full realization as a continuous complex $\mathfrak{K}$ of mesh $< \frac{1}{3}\epsilon$. Let ζ be the set of the faces of ω realized in $\mathfrak{K}_1$. The continuous cells making up the image of ω in $\mathfrak{K}$ are of diameter $< \frac{1}{3}\epsilon$ and each meets ζ. Since diam $\zeta <$ $\eta(\frac{1}{3}\epsilon, p) < \epsilon$, the image of ω is of diameter $< \omega$. Therefore $\mathfrak{K}$ considered as a full realization of Π is a continuous prism whose mesh $< \epsilon$. This proves both (8.4) and (8.5).

(8.6) Remark. In the sequel whenever prismatic realizations are envisaged the extension functions are assumed to be the prismatic extension functions. If $\eta(\epsilon, p)$ is such a function it may serve *a fortiori* as extension function for realizations of Euclidean complexes.

9. We shall now prove a convenient analogue of the deformation theorem of Alexander-Veblen for complexes (AT,VIII,23.6).

(9.1) DEFORMATION THEOREM. When $\mathfrak{R}$ is an LC^p compactum then corresponding to every $\epsilon > 0$ there exists in $\mathfrak{R}$ a finite continuous complex Ψ of dimension $\leq p$, and a $\xi(\epsilon) > 0$, such that every finite continuous complex $\mathfrak{K}$ in $\mathfrak{R}$ whose dimension $\leq p$ and mesh $< \xi(\epsilon)$ is ϵ homotopic to a subcomplex of Ψ. When $\mathfrak{R}$ is LC^* the same holds with all dimensional restrictions removed. When $\mathfrak{R}$ is LC^{pp} every $\mathfrak{K}$ whose dimension $\leq p$ is homotopic to a point, and when $\mathfrak{R}$ is LC** this holds for all $\mathfrak{K}$.

Let $\mathfrak{R}$ be LC^p with $\eta(\epsilon,p)$ as its extension function for continuous polyhedra (8.6). Take any $\epsilon > 0$, set $\zeta = \frac{1}{2}\eta(\frac{1}{3}\eta(\epsilon,p))$ and take a finite open irreducible covering $\mathfrak{U} = \{U_i\}$ such that $\Phi = \text{nerve } \mathfrak{U} \cong \text{nerve } \overline{\mathfrak{U}}$ (AT,VII,3.3). We may choose as the vertex A_i of Φ representing U_i a point which is in U_i but in no other set of $\mathfrak{U}$ (AT,VII,3.2). Denote by Φ^p the p-section of Φ, and by Φ_e, Φ_e^p Euclidean representatives of Φ, Φ^p.

Since mesh $\Phi^p < 2\zeta$, $\{A_i\}$ is a partial realization of mesh $< 2\zeta$ of Φ_e^p, and so it may be completed to a full realization $\Psi = (\Phi_e^p, t)$ of mesh $< \frac{1}{3}\eta(\epsilon,p)$. When $\mathfrak{R}$ is LC* we have instead a realization Ψ of Φ_e of mesh $< \frac{1}{3}\eta(\epsilon)$. At all events in both cases Ψ is simplicial. Let $\xi(\epsilon) = \inf \{\zeta, \text{Lebesgue number } \overline{\mathfrak{U}}\}$. Define also a transformation $s\colon \mathfrak{R} \longrightarrow \{A_i\}$ by the condition that $sx = A_i$ is such that $x \in U_i$.

Consider now a finite $\mathfrak{K} = (K, \theta)$ where $K = \{\sigma\}$, $\{B_i\}$ are the vertices of K, $\dim K \leq p$ and mesh $\mathfrak{K} < \xi$. If $\sigma = B_i \ldots B_j \in K$, then $\theta B_i, \ldots, \theta B_j$ are the vertices of (σ, θ) and so their set is of diameter $< \xi$. It follows that $s\theta B_i, \ldots, s\theta B_j$ are the vertices of a simplex of Φ^p (since $\dim \mathfrak{K} \leq p$) and consequently $s\theta$ defines a simplicial transformation $K \to \Phi^p$ and finally a barycentric mapping $\tilde{s}\colon |K| \longrightarrow |\Phi_e^p|$. Consider now the prism $1 \times K$, where 1 is the segment $0 \leq u \leq 1$, and let T be a mapping defined on $|L|$, $L = 0 \times K \cup 1 \times K$, as follows: if $x \in |K|$ then $T(0 \times x) = \theta x$, $T(1 \times x) = t\tilde{s}x$. We have thus in (L,T) a partial realization $\mathfrak{L}$ of $1 \times K$. Since each cell of $\mathfrak{L}$ meets one of a pair of points θB_i, $s\theta B_i$ or both, and since $d(\theta B_i, s\theta B_i) \leq \text{mesh } \mathfrak{U} < \zeta$, the mesh of $\mathfrak{L}$ (as a partial realization) $< \eta(\epsilon,p)$.

Therefore we may complete to a full realization of mesh $< \epsilon$ of $I \times K$ and this proves (9.1) for LC^p.

For the other cases the situation is as follows: (a) $\mathfrak{R}$ is LC^*. The proof is the same without dimensional restrictions. (b) $\mathfrak{R}$ is LC^{pp}. Then we may take $\mathfrak{U} = \mathfrak{R}$, hence Ψ = a point and carry out the proof without reference to meshes, thus leading to the corresponding part of (9.1). (c) When $\mathfrak{R}$ is LC^{**} both dimensional and mesh restrictions may be disregarded in the proof and as under (b) we may choose Ψ = a point, leading again to the corresponding part of (9.1). This completes the proof of the theorem.

(9.2) It may be pointed out that owing to the definition of s the deformation of (9.1) is uniquely defined for a given $\mathfrak{K}$.

10. Convex compacta provide the simplest examples of LC^* spaces. We shall need the

(10.1) DEFINITION. If R is convex and α is a subset of R, we understand by the <u>convex closure</u> of α, written $\widehat{\alpha}$, the intersection of all the closed convex sets of R containing α.

We shall now prove:

(10.2) Every convex compactum $\mathfrak{R}$ in an Euclidean or Hilbert parallelotope is LC^{**}.

Since R is contractible it is C^{ω} so we only need to prove it an LC^*. Let $\mathfrak{K}_1$ be a partial realization of $K = \{\sigma\}$ in R, with $\mathfrak{K}_1 = (D, t)$, L closed and dense in K. If σ_i^q is a simplex of K let $\zeta_i^q = t|L \cap \mathrm{Cl}\,\sigma_i^q|$. I say that we may insert the missing elements without modifying the $\widehat{\zeta}_i^q$. The zero-elements are already there so we may assume that the elements of dimension $< q$ have already been inserted in the required way. It follows that if σ^q has not yet been inserted then $|\mathfrak{B}\sigma^q|$ is already mapped as a subset of $\widehat{\zeta}^q$. By taking the join of a point of $\widehat{\zeta}^q$ with the points of $t|\mathfrak{B}\sigma^q|$ there is obtained a continuous q-cell in $\widehat{\zeta}^q$ which constitutes the extension of t to $\bar{\sigma}^q$, with $t\sigma^q \subset \widehat{\zeta}^q$. Therefore the insertion may be carried out as desired. Since diam $\widehat{\zeta}$ = diam ζ, the function $\eta(\epsilon) = \epsilon$ is an extension-function for R. Thus R is LC* also, and hence it is LC**.

(10.3) APPLICATION. Finite or Hilbert parallelotopes are LC**.

11. A useful property in connection with extending partial realizations is:

(11.1) Let K be a finite Euclidean complex with the closed subcomplex L and let $\mathfrak{R}_1' = (L, t_1')$ be a partial realization of K in $\mathcal{R}$. Suppose that t_1' is ϵ homotopic to $t_2' : L \longrightarrow \mathcal{R}$ thus yielding a partial realization $\mathfrak{R}_2' = (L, t_2')$. If $\mathfrak{R}_2'$ may be completed to a full realization $\mathfrak{R}_2$ of mesh $< \eta$ then $\mathfrak{R}_1'$ may be completed to a full realization of mesh $< 2\epsilon + \eta$. (Hurewicz [e]).

Let $l = \overline{ab}$ be a segment and let $\mathfrak{R}_2 = (K, t_2)$, $t_2' = t_2 \,|\, |L|$. By assumption there is a mapping $s: |l \times L \cup b \times K| \longrightarrow \mathcal{R}$ such that $s(a \times x) = t_1'x, x \in |L|$ and $s(b \times x') = t_2x'; x' \in |K|$. The problem is to extend s to $|l \times K|$.

Let $K = \{\sigma\}$. Since every σ^0 of K is in L, s is already defined on $l \times \sigma^0$. Assume it defined on all the $l \times \bar{\sigma}^q$ but not on a given $l \times \sigma^{q+1}$. At all events s is certainly defined on $|l \times \mathfrak{B}\sigma^{q+1} \cup b \times \sigma^{q+1}|$. If g is the centroid of σ^{q+1} take a point A in the space of $L \times \sigma^{q+1}$ on the line $1 \times g$ extended beyond $a \times g$. If P is a point of $1 \times \sigma^{q+1}$ the ray AP meets $|1 \times \mathcal{B}\sigma^{q+1} \cup b \times \sigma^{q+1}|$ in a point Q. Let t denote the mapping $1 \times \sigma^{q+1} \longrightarrow |1 \times \mathcal{B}\sigma^{q+1} \cup b \times \sigma^{q+1}|$ wherein $P \longrightarrow Q$. We define s on $1 \times \sigma^{q+1}$ by choosing $sP = sQ = stP$ and it is thus extended to $1 \times \sigma^{q+1}$. Hence s may be extended to $|1 \times K|$. The resulting realization completing $\mathfrak{K}_1'$ is the continuous complex $\mathfrak{K}_1 = (K, s_1)$ where $s_1 x = s(a \times x)$, $x \in |K|$.

The inductive construction of the images of the σ^q is such that $s_1\sigma^q$ consists of $t_2\sigma^q$ together with sets $s(1 \times \sigma^r)$, where $\sigma^r \prec \sigma^q$. Since diam $t_2\sigma^q < \eta$, and diam $s(1 \times \sigma^q) < \epsilon$, each point of $s(1 \times \sigma^r)$ is <u>nearer</u> than ϵ to some point of $s(b \times \sigma^r) \subset t_2\sigma^q$. Hence mesh $\mathfrak{K}_1 < 2\epsilon + \eta$ and (11.1) is proved.

(11.2) APPLICATION. Every finite Euclidean complex K is LC*.

Given any $\epsilon > 0$ choose a derived $K^{(n)}$ of K whose mesh $< \frac{\epsilon}{3}$. Let $\mathfrak{L}_1$ be a partial realization in K of a geometric complex L, where $\mathfrak{L}_1 = (L_1, t)$, L_1 closed and dense in L. Referring to the proof of (AT, VIII,23.6) we find that there is an $\eta(\epsilon) > 0$ such that if mesh $\mathfrak{L}_1 < \eta$ then $\mathfrak{L}_1$ is $\frac{\epsilon}{3}$ homotopic to a partial realization which is a subcomplex of $K^{(n)}$ and such that if σ is a non-realized face of L then $\mathcal{B}\sigma$ is imaged into a set of faces of a simplex of $K^{(n)}$. The extension may therefore be completed dimension by dimension by inserting simplexes of $K^{(n)}$. Since here the ϵ, η of (11.1) are both $\frac{\epsilon}{3}$, the full realization $\mathfrak{L}$ of L completing $\mathfrak{L}_1$ is of mesh $< \epsilon$. Therefore $\eta(\epsilon)$ is an extension function for K independent of dimensions and so K is LC*.

§3. RELATIONS BETWEEN THE LC PROPERTIES AND RETRACTION

12. We shall prove a basic theorem establishing, for compacta, the identity between the LC and absolute retract types. For the proof however, it is most convenient to obtain an expression of the LC properties in terms of regular infinite complexes. This is also interesting for its own sake since it will enable us to phrase the LC properties without any extension function whatever.

Before proceeding it will be convenient to extend the term "regular" to continuous and singular complexes. If $\mathfrak{K} = (K,t)$ is a continuous complex in the metric space $\mathfrak{R}$, then $\mathfrak{K}$ is said to be regular in $\mathfrak{R}$ whenever: (a) K is Euclidean; (b) if $K = \{\sigma_n\}$ then diam $t\sigma_n \longrightarrow 0$. Similarly if A is a closed subset of $\mathfrak{R}$ then $\mathfrak{K}$ is regular relative to A whenever (a) holds and in addition (b) is strengthened to: (b') sup $\{\text{diam } t\sigma_n, d(t\sigma_n, A)\} \longrightarrow 0$. If $\mathfrak{K}$ is a singular complex in $\mathfrak{R}$ then $\mathfrak{K}$ is regular [regular relative to A] whenever: (a) $\mathfrak{K}$ is countable locally-finite; (b) if $\mathfrak{K} = \{E_n\}$ then diam $E_n \longrightarrow 0$ [sup$\{\text{diam } E_n, d(|E_n|, A)\}\longrightarrow 0$.

Let K, L, $\mathfrak{K}_1 = (L,t)$ be as in (5) save that K is merely an Euclidean complex and in particular countable and locally finite. If $K = \{\sigma_n\}$ we set $\zeta_n = t|L \cap \text{Cl}\sigma_n|$. Then if $\{\zeta_n\}$ is regular we say that the partial realization is regular. Let K_0 be closed and dense in K and let its complement be finite. Then a regular realization of K_0 is said to be an almost complete regular realization of K.

(12.1) DEFINITION. The compactum $\mathfrak{R}$ is said to be $\widetilde{LC}^p$ whenever every regular partial realization of an Euclidean complex K whose dimension $\leq p+1$ may be extended to an almost complete regular realization. If the same property holds whatever the dimension of K then $\mathfrak{R}$ is said to be $\widetilde{LC}$*. If the realizations may be made complete $\mathfrak{R}$ is said to be $\widetilde{LC}^{pp}$ in the first case, $\widetilde{LC}$** in the second.

(12.2) THEOREM. For a compactum $\mathfrak{R}$ the following equivalences take place between the local connectedness and retraction properties:

(12.2a) $ANR \Longleftrightarrow LC^* \Longleftrightarrow \widetilde{LC}^*$

(12.2b) $AR \Longleftrightarrow LC^{**} \Longleftrightarrow \widetilde{LC}^{**}$

(12.2c) $ANR^{p+1} \Longleftrightarrow LC^p \Longleftrightarrow \widetilde{LC}^p$

(12.2d) $AR^{p+1} \Longleftrightarrow LC^{pp} \Longleftrightarrow \widetilde{LC}^{pp}$.

A noteworthy consequence of (12.2) is:

(12.3) For compacta corresponding LC and $\widetilde{LC}$ properties are equivalent. That is to say the definitions by means of finite and those by means of infinite partial realizations are equivalent.

From (12.2a) and (4.3), also from (12.2b) and (2.4) we obtain:

(12.4) Every LC* compactum is locally contractible and every LC** compactum is contractible.

(12.5) Finite polyhedra are LC*. (Second proof) (III,18.4; 12.2a).

(12.6) <u>Notations</u>. P^ω, P_1^ω are as in (III,17.4). The basic compactum R is chosen in P_1^ω and there is constructed in $P^\omega - R$ the complex of (III,19.1); it is denoted this time by H and its n-section by H^n. We designate by ρ the retraction $P^\omega \to R \cup |H|$ of (III,19.1) with the basic property: $\rho(P^\omega - R) = |H|$.

(12.7) Method of proof. To establish (12.2a) it is sufficient to prove the circular implication ANR $\Longrightarrow$ LC* $\Longrightarrow$ $\widetilde{\text{LC}}$* $\Longrightarrow$ ANR. The three successive implications ANR $\Longrightarrow$ LC*, ... will be denoted by (A*), (B*), (C*) and the corresponding implications for (12.2 bcd) by (A**), ..., (A^p), ... (A^{pp}), ...

13. (13.1) Proof of (A*): ANR $\Longrightarrow$ LC*. Let R be an ANR. There exists then a neighborhood N of R in P^ω with a retraction $\theta : N \to R$. We shall need:

> (13.2) Given any $\epsilon > 0$ there is an $\eta(\epsilon) > 0$ such that if diam $\alpha < \eta$ and α meets R then diam $\theta\alpha < \epsilon$. (AT,I,45.7).

We shall now show that $\eta(\epsilon)$ is an extension function for R. Let in fact K, $\mathfrak{R}_1$, L, t be as in (5) where as in (5) $\mathfrak{R}_1 = (L,t)$ is a partial realization of K of mesh $< \epsilon$ (as a partial realization). Since P is convex we may proceed as in the proof of (10.2) and obtain a complete realization $\mathfrak{R}_2 = (K, t_1)$ of mesh $< \eta$ in P^ω. Then $\mathfrak{R} = (K, \theta t_1)$ is a complete realization in R of mesh $< \epsilon$. Since $t_1 \mid L = t$, and $\theta t_1 \mid L = t_1 \mid L = t$, $\mathfrak{R}$ extends $\mathfrak{R}_1$. Therefore $\eta(\epsilon)$ is an extension function, R is LC* and (13.1) is proved.

(13.3) Proof of (A**). If R is an AR it is also an ANR and hence already an LC* by (A*). Let σ be any Euclidean simplex and t a mapping $|\mathfrak{B}\sigma| \to R$. Since R is an AR we may extend t to σ and so R is C^ω. Therefore it is an LC** and this is (A**).

(13.4) Proof of (A^p). The proof will rest upon the following property:

(13.5) If R is an ANR^{p+1} then given any $\epsilon > 0$ there is an $\eta > 0$ such that if (K,t) is a countable locally finite continuous complex, regular relative to R, of dimension $< p + 1$ and in $\mathfrak{S}(R,\eta) - R$ then there is an ϵ mapping $\tau: \overline{t|K|} \longrightarrow R$ which is the identity on $t\,|\,K\,| \cap R$.

Since R is an ANR^{p+1} it has a neighborhood N_1 in $R \cap H^{p+1}$ (see 12.6) with a retraction $\rho': N_1 \longrightarrow R$. There exists then a neighborhood N_2 of R in $R \cup |H|$ such that if A is the union of the closed simplexes of H which meet N_2 then the $(p+1)$-section B of A is in N_1. Hence $N = \rho^{-1} N_2$ (ρ as in 12.6) is a neighborhood of R in P^{ω}. If $x \in N - R$ then ρx is on a $\sigma \in A$ and so if y_x is any point of a face σ^r of σ, $r \leq p + 1$, we will have $d(\rho x, y_x) < \text{diam } \sigma$. Since $y_x \in |B|$, we have $\rho' y_x \in R$. Since H is regular relative to R $\rho| R = 1$, we may choose N_2, hence N such that whatever y_x (chosen as above), then whatever $x \in N - R$ we have $d(x, \rho' y_x) < \epsilon$. Moreover as $x \longrightarrow R$ we have $d(x,y_x) \longrightarrow 0$. Finally we choose η such that $\mathfrak{S}(R,\eta) \subset N$.

Let now (K,t) be as in (13.5). Then $(K,\rho t)$ is a continuous complex in $|A|$ and still regular relative to R. It follows from this that if $\sigma \in K$ then $d(\rho t\sigma, R) > 0$. In view of the regularity of H relative to R, each element of one of H and $(K,\rho t)$ meets at most a finite number of elements of the other. We may therefore apply (AT,VIII,23.7) thus obtaining an image $(K, t'\rho t)$ of K in $|B|$ and $(K,\rho' t'\rho t)$ will be an image in R. If we set $\tau = \rho' t' \rho$, we find that, owing to the choice of η. τ behaves as required in (13.4), which is thus proved.

The application to the proof of (A^p) is immediate. Suppose that for $r \leq p$ we have a mapping $t: |\mathfrak{B}\,\sigma^{r+1}| \longrightarrow R$ such that $\text{diam } t|\mathfrak{B}\,\sigma^{r+1}| < \epsilon$. Applying now (AT,VIII,5.1) we replace σ^{r+1} in $\bar{\sigma}^{r+1}$ by a polyhedron $|K|$, where K is an Euclidean complex regular relative to $|\mathfrak{B}\,\sigma^{r+1}|$. Then following (III,24.2) we may extend t to a mapping $t_1 : \bar{\sigma}^{r+1} \longrightarrow P^{\omega}$, taking care to choose the images of the vertices of K in the intersection C of the convex closure of $t|\mathfrak{B}\,\sigma^{r+1}|$ with $P^{\omega} - P_1^{\omega}$. As a consequence we will also have $(K, t_1) \subset C$. Then we have τ of (13.4) and τt_1 will be an extension of t to an $\frac{\epsilon}{3}$ mapping $\bar{\sigma}^{r+1} \longrightarrow R$, and diam $\tau\sigma^{r+1} < \epsilon$. Therefore R is r - LC for every $r \leq p$ and so it is LC^{p}.

(13.6) <u>Proof of (A^{pp})</u>. Essentially as in (13.3) save that dim $\sigma \leq p+1$.

14. (14.1) <u>Proof of (B^*) : $LC^* \Longrightarrow \widetilde{LC}^*$</u>. Suppose that R is LC^* with the extension function $\eta(\epsilon)$. Let K, L, $\mathfrak{K}_1$, t be as in (5) save that K is any Euclidean complex and the partial realization $\mathfrak{K}_1$ is regular (12). Choose now a sequence $\{\epsilon_i\} \longrightarrow 0$ monotonely and set $\eta_i = \eta(\frac{1}{3}\eta(\epsilon_i))$. If $K = \{\sigma\}$ we define $\lambda(\sigma) =$ diam $t|L \cap \mathrm{Cl}\,\sigma|$. Notice at the outset that if $\lambda(\sigma) = 0$ we may assume σ and all its faces to be in L. For σ has a vertex A in L and we may then extend t to σ by defining $t\sigma = tA$.

We will construct now a sequence of finite subcomplexes $P_1, Q_1, R_1, P_2, \ldots$ of K where the P_i are open and the others are closed. P_1 is an open finite subcomplex which includes every σ such that $\lambda(\sigma) \geq \eta_1$. Since K is regular the number of such σ is finite and so P_1 may always be constructed. For the same reason we may take for Q_1 a closed finite subcomplex of $K - P_1$, hence also of K, which includes every σ of $K - P$ such that $\eta_2 \leq \lambda(\sigma) < \eta_1$. Then we take $R_1 = P_1 \cup Q_1$ and it is a closed finite subcomplex of K which includes every σ such that $\lambda(\sigma) \geq \eta_2$. Suppose that we have already constructed $P_1, \ldots, R_{i-1}$ such that $R_{i-1} = P_1 \cup \ldots \cup Q_{i-1}$, is finite and closed in K, and includes every σ such that $\lambda(\sigma) \geq \eta_i$. Define now

$P_i = \mathrm{St}\, R_{i-1} - R_{i-1}$, Q_i = a closed subcomplex of $K - \mathrm{St}\, R_{i-1}$ (hence also of K) which includes every σ not in $K - \mathrm{St}\, R_{i-1}$ such that $\eta_{i+1} \leqq \lambda(\sigma) < \eta_i$. Then clearly $R_i = P_1 \cup \ldots \cup Q_i = R_{i-1} \cup P_i \cup Q_i$ is finite and closed in K and includes every σ such that $\lambda(\sigma) \geqq \eta_{i+1}$. The inductive construction is thus completed.

The sequence $P_1, Q_1 P_2, Q_2, \ldots$ is such that: (a) the Q_i are disjoint finite closed subcomplexes of K; (b) if $\sigma \in P_i$, $i > 1$, then $\sigma = \sigma'\sigma''$, $\sigma' \in Q_{i-1}, \sigma'' \in Q_i$; (c) Q_i and $\mathrm{Cl}\, P_{i+1}$ have partial realizations of mesh $< \eta_i$.

It is a consequence of (c) that the partial realization of Q_i may be extended to a full realization Q_i^* whos mesh $< \frac{1}{3}\eta(\epsilon_i)$. From (c) follows then that $\mathrm{Cl}\, P_i$, $i > 1$, has then a partial realization of mesh $< \eta(\epsilon_{i-1})$ and so it may be extended to a-full realization which contains one P_i^* of P_i to mesh $< \epsilon_{i-1}$. We obtain thus a regular realization of $K - P_1$, that is to say an almost complete regular realization of K. Therefore R is $\widetilde{LC}^*$ or (B*) holds.

(14.2) <u>Proof of (B**)</u>. When R is LC^{**} it is also C^ω. Hence the missing elements of P_1 in the almost complete realization may be inserted dimension by dimension, thus yielding a complete realization. Therefore R is LC^{**} which is (B**).

(14.3) <u>Proof of (B^p) and (B^{pp})</u>. The same as for (B*) <u>and (B**) save that dim</u> $K \leqq \underset{\sim}{p} + 1$.

15. (15.1) <u>Proof of (C*): $\widetilde{LC}^* \Longrightarrow$ ANR.</u> Let R be LC^*, and let $\{A_i\}$ be the vertices of H of (12.6). Corresponding to each A_i select a point B_i in R such that $d(A_i, B_i) = d(A_i, R)$. Since H is regular relative to R the set $\{B_i\}$ is a regular partial realization of H and so it may be extended to an almost complete regular realization $\{H_1, t\}$, where H_1 is a closed subcomplex of H such that $H - H_1$ is finite.

Since the realization is regular t may be extended to a mapping $t_1 : |H_1| \cup R \longrightarrow R$ such that $t_1|R$ is the identity. There exists also a neighborhood N_1 of R in $R \cup |H|$ such that $N_1 \subset R \cup |H_1|$. Therefore $N = \rho^{-1}N_1$ is retracted by $t_1\rho$ into R, or R is an ANR.

(15.2) <u>Proof of (C**)</u>. This time we may choose $H_1 = H$, hence $N = P^\omega$ and so R is an AR.

(15.3) <u>Proof of (C^p)</u>. Let R be LC^p and let it have a topological image R_1 in a separable metric space S_1 such that $\dim(S_1 - R_1) \leqq p + 1$. Let also t be a topological mapping $R_1 \longrightarrow R$. Since $R \cup |H^{p+1}|$ is an ANR^{p+1} we may extend t to a mapping $t_1 : S_1 \longrightarrow R \cup |H^{p+1}|$. By the same reasoning as in (15.1) there is a neighborhood N of R in $R \cup |H^{p+1}|$ with a retraction $\tau : N \longrightarrow R$. Then $N_1 = t_1^{-1} N$ is a neighborhood of R_1 in S_1 mapped by τt_1 into R where $\tau t_1 | R_1 = t$. Therefore $t^{-1}\tau t_1$ is a retraction $N_1 \longrightarrow R_1$ and so R_1, hence also R, is an ANR^{p+1}.

(15.4) <u>Proof of C^{pp}</u>). When R is LC^{pp} we may choose $N = R \cup |H^{p+1}|$, hence $N_1 = S_1$ and so $t^{-1}\tau t_1$ is a retraction $S_1 \longrightarrow R_1$. Hence R_1, and also R, is an AR^{p+1}.

§4. CHARACTERIZATION OF THE LC PROPERTIES BY MAPPINGS ON CONTINUOUS COMPLEXES

16. (16.1) DEFINITION. Let $\mathfrak{R}$, A be topological spaces and $\mathfrak{K} = (K, t)$ a continuous complex on $\mathfrak{R}$. A mapping $s : A \longrightarrow \mathfrak{R}$ will be called a <u>mapping into the continuous</u> complex $\mathfrak{K}$ whenever there exists a mapping $t_1 : A \longrightarrow |K|$ such that $s = tt_1$.

We prove several propositions which show the relations existing between mappings into continuous complexes and the LC properties.

(16.2) THEOREM. A n.a.s.c. for a compactum $\mathfrak{R}$ to be LC* is that $\mathfrak{R}$ be ϵ deformable whatever ϵ into a finite continuous complex. (Lefschetz [f]).

(16.3) THEOREM. A n.a.s.c. for a compactum $\mathfrak{R}$ to be LC^p is that given $\epsilon > 0$ there exists a finite continuous p-complex $\mathfrak{K}^p$ such that every mapping s of a compactum F into $\mathfrak{R}$, where $\dim F \leq p$, is ϵ homotopic to a mapping of F into $\mathfrak{K}^p$.

For LC* and LC^{pp} spaces we have:

(16.4) $LC^{**} \Longleftrightarrow LC^{*}$ + contractibility.

(16.5) $LC^{pp} \Longleftrightarrow LC^p$ + every mapping of an at most p-dimensional compactum into the space is homotopic to a point.

These propositions will be proved in the order: (16.2), (16.4), (16.3), (16.5).

17. Proof of (16.2). Let $\mathfrak{R}$ be an LC* compactum, R a topological image of $\mathfrak{R}$ in P^ω, and N, e, ϵ, η as in (13.1). We impose in addition $\eta < \inf \{\epsilon, d(R, P^\omega - N)\}$. Choose now a finite $\eta/3$ open covering $\{U_i\}$ of R by open sets of P^ω and construct an Euclidean nerve K of the covering in P^ω with the vertex A_i representing U_i a point of U_i itself. Referring to (AT,VIII,3.4) it will be seen that the construction of K in P^ω is always possible. If A_iA_j is a simplex of K then $U_i \cap U_j \neq \emptyset$, hence $\operatorname{diam} A_iA_j < \frac{2}{3}\eta$ (AT,VIII,2.3). Let τ be the canonical mapping $R \longrightarrow |K|$. Since $x \in U_i \Longrightarrow \tau x \in |\mathrm{Cl\ St}\ A_i|$, there follows $d(x, \tau x) < d(x, A_i) + \operatorname{mesh} K < \eta$. Hence the segment $\lambda(x)$ joining x to τx is in N and its length $< \eta$.

Let $\Lambda = \bigcup \lambda(x)$. It is clear that τ is an η deformation of R into K over $\bar{\Lambda}$ with the $\lambda(x)$ as the paths (AT,I,47.4). Therefore by (13.2) $\theta\tau$ is an ϵ deformation of R over itself into the continuous complex $\mathfrak{K} = (K, \theta)$. This proves the necessity of the LC* condition.

Suppose now the condition of (16.2) fulfilled for $\mathcal{R}$. There exists for given $\epsilon > 0$ an $\epsilon/3$ deformation s of $\mathcal{R}$ into a finite continuous complex $\mathfrak{K} = (K,t)$, where $s = tt_1$, t_1 a mapping $\mathcal{R} \longrightarrow |K|$. Since $|K|$ is LC* (11.2) it has an extension function $\xi(\epsilon)$. Since t is continuous and $|K|$ is a compactum there is a $\zeta > 0$ such that if $\alpha \subset |K|$ and diam $\alpha < \zeta$ then diam $(t\alpha) < \frac{\epsilon}{3}$. Since t_1 is continuous and $\mathcal{R}$ is a compactum there exists also an $\eta > 0$ such that if $\beta \subset \mathcal{R}$ and diam $\beta < \eta$ then diam $(t_1 \beta) < \xi(\zeta)$.

Let now the finite Euclidean complex L have in $\mathcal{R}$ a partial realization (L_1, τ_1) whose mesh $< \eta$. Then $(L_1, t_1\tau_1)$ is a partial realization in $|K|$ whose mesh $< \xi(\zeta)$ and so it may be extended to a full realization (L, τ) of mesh $< \zeta$ in $|K|$. Hence $(L, t\tau)$ is a full realization in $\mathcal{R}$ which extends $(L_1, s\tau_1)$ and whose mesh $< \frac{\epsilon}{3}$. Therefore by (11.1, the partial realization (L_1, τ_1) may be extended to a full realization of mesh $< \epsilon$. Thus $\eta(\epsilon)$ is an extension function for $\mathcal{R}$ independent of dim L, and so $\mathcal{R}$ is LC*. This proves sufficiency and hence (16.2).

Proof of (16.4). If $\mathcal{R}$ is LC** it is LC* and also an AR (12.2b) hence contractible. Conversely if $\mathcal{R}$ is contractible it is C^ω, and if it is also LC* it is then LC**. This proves (16.4).

18. Proof of (16.3). Let the notations be those of (12.5) with R an LC^p compactum and in P_1^ω. Take a finite open irreducible α covering $\mathfrak{U}^*$ of P^ω, such that nerve $\mathfrak{U}^*$ = nerve $\overline{\mathfrak{U}}^*$ (AT,VII,3.3). The constant α will be specified later. We set $\beta = \inf \{\alpha$, Lebesgue number $\overline{\mathfrak{U}}^*\}$ and denote by $\mathfrak{U} = \{U_i\}$ the sub-

collection of $\mathfrak{U}^*$ which covers R. It is in fact only with $\mathfrak{U}$ that we shall be dealing hereafter.

Construct now as in (17) an Euclidean nerve Ψ for $\mathfrak{U}$ by choosing as the vertex representing U_i a point $C_i \in U_i \cap (P^\omega - P_1^\omega)$. Since $P^\omega - P_1^\omega$ is convex it will contain Ψ. We notice also that if $C_i C_j$ is a simplex of Ψ then $U_i \cap U_j \neq \emptyset$, hence $d(C_i, C_j) < 2\alpha$ and so mesh $\Psi < 2\alpha$.

Let now t be a mapping $F \longrightarrow R$, where F is a compactum whose dimension $\leq p$. By (III,16.2) applied to F we may suppose $F \subset G \subset P^\omega$, where G is a compactum such that $G - F = |M|$ has the property of (III,16.2) with n replaced by p. Applying (III,24.3) t may be extended to a mapping $t_1 : G \longrightarrow P^\omega$ such that $t_1 \mid |M|$ is topological while $|L| = t_1|M| \subset (\bigcup U_i) \cap (P^\omega - P_1^\omega) \subset P^\omega \dot{-} R$, L being an Euclidean complex regular relative to R and of mesh $< \beta$. Furthermore referring to (III,24.3) the construction may be so carried out that any q vertices of $\Psi \cup L$ have a coordinate matrix of rank q.

It is now a consequence of (III,16.2) and the continuity of t_1, that L contains a closed subcomplex L_1 and L_1 a finite closed subcomplex Φ whose dimension $\leq p$, such that $t : F \longrightarrow R$ is homotopic over $R \cup |L_1|$ to a mapping $t' : F \longrightarrow |\Phi|$.

Let $\{A_i\}$ be the vertices of Φ. To each A_i assign a vertex C_j of Ψ such that $A_i \in U_j$. Since mesh $\Phi \leq$ mesh $L < \beta$, $A_i \longrightarrow C_j$ defines a barycentric mapping $\tau : |\Phi| \longrightarrow |\Psi|$, and in fact $|\Phi| \longrightarrow |\Psi^p|$, since dim $\Phi \leq p$. Proceeding as in (AT,VIII,23.5) τ is turned into a deformation over $|Q|$, Q a finite Euclidean complex whose simplexes are simplexes of Φ, Ψ^p or else of the form $\sigma'\sigma''$, $\sigma' \in \Phi$, $\sigma'' \in \Psi^p$. Furthermore since any q vertices of $L \cup \Psi$ are the vertices of a σ^{q-1},

$L \cup Q$ is an Euclidean complex. An estimate of mesh Q is required. Its one-simplexes are segments whose vertices are among those of a $\sigma' \in \Phi$ and those of the corresponding $\tau\sigma' = \sigma''$. That is to say they are of the forms $A_1A_h \in \sigma'$, $C_1C_h \in \sigma''$, and $A_1(\tau A_h)$ where $A_1A_h \in \sigma'$, whose lengths are respectively $< \beta$, 2α, α. Hence mesh $Q < 2\alpha$. Since the path of $x \in \sigma'$ under the deformation τ is in a star of a vertex of Q, its diameter $<$ 2 mesh $Q < 4\alpha$. Hence τ is a 4α deformation. Therefore t' is 4α homotopic to $t_0 = \tau t' : F \longrightarrow |\Psi^p|$ over $|Q|$. It follows that t is 5α homotopic over $R \cup |L| \cup |Q|$ to the mapping t_0.

We will now choose $\alpha < \eta/5$ and set $L \cup Q \cup \Psi^p = \Lambda$. The latter is an Euclidean complex, of dimension $\leq p + 1$, regular relative to R, contained in $\mathfrak{S}(R, \eta) - R$, and such that t is η homotopic over $\bar{\Lambda}$ to a mapping t_0: $F \longrightarrow |\Psi^p|$.

Now the closed simplexes of H' which meet $\rho|\Psi^p|$ (notations of 12.5) make up a finite closed subcomplex of H whose p-section is denoted by K^p. Referring to (13.4) there is an ϵ mapping $\phi : |\bar{\Lambda}| \rightarrow R$ which is the identity on $R \cap |\bar{\Lambda}|$ and t_0 is a mapping $F \rightarrow \mathfrak{K}^p = (K^p, \rho')(\rho'$ as in the proof of 13.4). Since $\phi t = t$ and t is homotopic to t_0 over $|\bar{\Lambda}|$, t is ϵ homotopic to ϕt_0 over R. This proves necessity for (16.3). Sufficiency is shown as for (16.2) save that the dimensions involved are $\leq p + 1$.

<u>Proof of (16.5)</u>. When $\mathfrak{R}$ is LC^{pp} the deformation just obtained is valid without any mesh restriction. Therefore we may choose for Ψ the nerve of the open covering of R by P^ω itself. Then Ψ is a point, and so the homotopy of F is to a point. Thus the left side in (16.4) implies the right side. Conversely when the conditions at the right hold $\mathfrak{R}$ is LC^p and C^p and hence it is LC^{pp}, proving (16.5).

19. Applications. (19.1) When a compactum $\mathfrak{R}$ is LC^p and its dimension $\leq p$ it is LC^*.

For in (16.2) we may choose $F = \mathfrak{R}$ and $s =$ the identity on $\mathfrak{R}$. As a consequence the condition of (16.1) is fulfilled and $\mathfrak{R}$ is LC^*.

(19.2) When a compactum is finite dimensional and locally contractible it is LC^*.

Let $\dim \mathfrak{R} = p$. Since $\mathfrak{R}$ is locally contractible it is LC^ω, hence a fortiori LC^p and therefore LC^* by (19.1).

The result just proved may be considered as a partial converse of (12.4) for LC^*.

(19.3) Finite geometric complexes are LC^*. (Third proof) (4.7, 19.2).

For the other two proofs see (11.2, 12.5).

(19.4) Second proof of the deformation theorem (9.1). An outline for LC^p will suffice. The situation being as in (16.5) let $\mathfrak{K}^p = (K^p, \theta)$, and let $\mathfrak{L} = (L, \tau)$, $\dim \mathfrak{L} \leq p$, be a continuous complex in $\mathfrak{R}$. By (16.3) τ is ϵ homotopic to a mapping $\tau_1 : |L| \longrightarrow \mathfrak{K}^p$. That is to say $\tau_1 = \theta t_1$ where t_1 is a mapping $|L| \longrightarrow |K^p|$. If $\mathfrak{L}$ hence L is of suitably small mesh, (L, t_1) is homotopic with a sub-complex (L, t_2) of K^p, and hence $\mathfrak{L}$ is homotopic with the sub-complex $(L, \theta t_2)$ of $\mathfrak{K}^p$ in $\mathfrak{R}$. By choosing mesh $\mathfrak{K}^p$, hence mesh K^{p} small enough this may be made a 2ϵ homotopy and this is in substance (9.1) for LC^p. The modifications for the other cases are obvious.

§5. HOMOLOGY THEORY OF LC SPACES

20. The ground has been so thoroughly prepared in (II) that the treatment of these questions will be rapid. In substance we shall find that LC* spaces have the properties of quasi-complexes in the sense of (AT,VIII, 31). In fact they are certainly quasi-complexes but the proof will not be given here. Moreover "quasi-complexes" only include LC* spaces; for the other categories and the similar comparison they should be duly generalized with some dimensional restriction on the cycles (in the sense of "quasi-complex" for dimensions $\leq$ p), but we shall not go into this question here.

21. Let $\mathcal{R}$ be a compactum, $\Omega = \{\sigma\}$ and $\Sigma = \{E\}$ its complete Vietoris and singular complexes, Ω^q and Σ^q their q-sections, (t,τ) the adherence $\Sigma \to \Omega$ (II,2.2). It will be recalled that tE is merely the simplex of Ω with the same vertices as E, and is a carrier for τ.

(21.1) When $\mathcal{R}$ is an LC^p compactum, there can be found essential subcomplexes Σ_1^{p+1}, $\Sigma_2^p \subset \Sigma_1^{p+1}$, Ω_1^{p+1} of Σ^{p+1}, Σ^p, Ω^p and a simplicial singular complex $\tilde{\Sigma}^p$ with the following properties:

(a) $t\,\Sigma_1^{p+1} = \Omega_1^{p+1}$;

(b) $\tau | \tilde{\Sigma}^{p+1} = \mu$ is a metric isomorphism
$$\tilde{\Sigma}^{p+1} \longrightarrow \Omega_1^{p+1};$$

(c) $\mu^{-1}(\tau | \Sigma_1^{p+1}) = \rho$ is a metric chain-mapping
$$\Sigma_1^{p+1} \longrightarrow \tilde{\Sigma}^{p+1};$$

(d) $\rho | \Sigma_2^p = \nu$ is a metric singular deformation
$$\Sigma_2^p \longrightarrow \tilde{\Sigma}^{p+1}$$

(21.2) Let P_q denote the property of the statement with q instead of p, and ν merely singular prismatic. Referring to (II, 6.3) it is sufficient to prove P_p. Let the statement " $\mathfrak{R}$ is an LC^{-1} compactum" merely mean " $\mathfrak{R}$ is a compactum", and let the related Σ_2^{-1}, stand for a null-complex. Then P_{-1} is trivially verified. Thus we merely need to assume P_{q-1}, $0 \leqq q \leqq p$, and derive P_q. We notice that (21.1c) is a consequence of (21.1a). Furthermore if (21.1b) holds we may take $\Sigma_1^{p+1} = t^{-1} \Omega_1^{p+1}$, and so (21.1a) will hold. Thus we only need to extend (21.1 bd). We are given the analogues $\Sigma_1^q, \ldots,$ of $\Sigma_1^{p+1}, \ldots,$ and we must find $\Sigma_1^{q+1}, \ldots$.

(21.3) By hypothesis we already have a metric isomorphism: $\mu : \tilde{\Sigma}^q \longrightarrow \Omega_1^q$. Hence if $\sigma^q \in \Omega_1^q$ then $\mu^{-1}\sigma^q$ exists and its diameter $\rightarrow 0$ uniformly with diam σ^q. It follows that if $\sigma^{q+1} \in \Omega$ and $\mathfrak{B}\sigma_1^{q+1} \in \Omega_1^q$ then $\mu^{-1}\mathfrak{B}\,\sigma^q$ exists and its diameter $\rightarrow 0$ uniformly with diam σ^{q+1}. Let $\eta(\epsilon, p)$ be a polyhedral extension function for $\mathfrak{R}$. Choose any $\epsilon > 0$ and let $\mu^{-1}\mathfrak{B}\,\sigma^{q+1} \subset \tilde{\Sigma}^q$, diam $|\mu^{-1}\mathfrak{B}\,\sigma^{q+1}| < \eta(\frac{\epsilon}{2n}, p)$. There exists an Euclidean simplex σ_e^{q+1} and a mapping $r : |\mathfrak{B}\,\sigma_e^{q+1}| \longrightarrow \mathfrak{R}$ such that $(\mathfrak{B}\,\sigma_e^{q+1}, r) = \mu^{-1}\mathfrak{B}\,\sigma^{q+1}$. Under our assumptions we may extend r to σ_e^{q+1} yielding $E^{q+1} = (\sigma_e^{q+1}, r)$ of diameter $< \epsilon/2^n$. We now add E^{q+1} to $\tilde{\Sigma}^q$ and extend μ to this new singular cell by defining $\mu E^{q+1} = \sigma^{q+1}$. This will manifestly yield a pair $\tilde{\Sigma}^{q+1}$, Ω_1^{q+1} related as required by P_q.

(21.4) For entirely similar reasons if $E^q = (\sigma^q, r)$ is such that $(\mathfrak{B}\,\sigma_e^{q+1}, r) \subset \Sigma_2^{q-1}$ then there is a mapping $S : |l \times \mathfrak{B}\,\sigma_e^q| \longrightarrow \mathfrak{R}$ (l is the segment $0 \leqq u \leqq 1$) which yields all the DE^r for every $E^r \prec E^q$, $r < q$, and is such that $S(0 \times \sigma_e^q) = E^q$, $S(1 \times \sigma_e^q) = tE^q$. Furthermore since μ is a metric chain-deformation $\Sigma_2^{q-1} \longrightarrow \tilde{\Sigma}^q$ we have at once that diam $S|\mathfrak{B}\,\sigma_e^q| \longrightarrow 0$

uniformly with diam E^q. Proceeding now as in (21.3) we may for a suitable $\Omega_2^q \subset \Omega_1^{q+1}$ extend S to σ_e^q and obtain $\mathfrak{D}E^q$ such that diam $(\mathfrak{D}E^q) \longrightarrow 0$ uniformly with diam E^q. As a consequence μ is a deformation such as required by P_q. Thus P_q holds and (21.1) follows.

(21.5) If $\mathfrak{R}$ is LC^p then every class of finite singular cycles Γ_s^q, $q \leq p$, contains simplicial cycles, hence also continuous cycles (21.1 cd), whose mesh is arbitrarily small.

By (II, 7.1) Γ_s^q contains a cycle γ^q of arbitrarily small mesh. We may therefore choose $\gamma^q \subset \Sigma_2^p$ and hence $\nu\gamma^q \subset \tilde{\Sigma}^{p+1}$ will be simplicial. Since ν is a singular metric deformation $\gamma^q \sim \nu\gamma^q$ and mesh $\nu\gamma^q \longrightarrow 0$ with mesh γ^q, (21.5) follows.

(21.6) Let now Ψ, ϵ, ξ be as in (9.1). It will be seen immediately that no part of the argument in (9) requires that the continuous complex $\mathfrak{C}$ be finite. Consider in particular $\tilde{\Sigma}^p$, the p-section of $\tilde{\Sigma}^{p+1}$. If we replace ϵ in the construction of $\tilde{\Sigma}^{p+1}$ in (21.3) by ξ, we will have mesh $\tilde{\Sigma}^p \leq$ mesh $\tilde{\Sigma}^{p+1} < \xi$. Since $\tilde{\Sigma}^p$ is simplicial it may be identified with a continuous complex and (9.1) applies to it. There results a prismatic singular deformation $\tilde{\Sigma}^p \longrightarrow \Psi$ which in turn may be replaced by an ordinary singular deformation $\theta : \tilde{\Sigma}^p \longrightarrow \Psi$ (II, 6.3).

On the other hand we may clearly choose Σ_2^p of (21.1) of arbitrarily small mesh, and in particular such that $\nu\Sigma_2^p \subset \tilde{\Sigma}^p$. Consequently $\theta\nu$ is a singular deformation $\Sigma_2^p \longrightarrow \Psi$.

When $\mathfrak{R}$ is LC* the situation is the same save that p may assume any finite value. Hence we have:

(21.7) When $\mathfrak{R}$ is an LC^p [LC*] compactum then corresponding to every $\epsilon > 0$ there is a finite simplicial complex Ψ, and for every $q \leq p$ [for every q] a number $\zeta(\epsilon,q)$ such that every singular complex $\mathfrak{C}$ whose dimension $\leq q$ and mesh $< \zeta(\epsilon,q)$ is ϵ singularly deformable into a subcomplex of Ψ. Moreover the deformation is uniquely determined for all $\mathfrak{C}$.

This result will be most convenient in connection with the fixed point theorem (25.4).

22. The ground is now prepared for the homology theory. We prove

(22.1) THEOREM. When $\mathfrak{R}$ is an LC^p [an LC^ω and a fortiori an LC*] compactum then its discrete Čech and Vietoris homology groups and the corresponding groups of the finite singular cycles for dimensions $q \leq p$ [for all dimensions] are isomorphic.

It is manifestly sufficient to treat the LC^p case. Moreover owing to (AT,VII,26.1) only the singular and Vietoris groups need be considered. Finally in view of (II, 7.3) the singular groups may be replaced by the VS-groups. Thus if $\mathfrak{H}_v$, $\mathfrak{H}_{vs}$ denote the Vietoris and VS-groups, we must prove

(22.2) $\mathfrak{H}^q_{vs}(\mathfrak{R},G) = \mathfrak{H}^q_v(\mathfrak{R},G)$, $q \leq p$.

Let the Vietoris and VS-classes be denoted by Γ^q_v and Γ^q_{vs} and otherwise let the notations be those of (21.1). The chain-adherence τ is already known to induce a homomorphism $\tau_1 : \mathfrak{H}^q_{vs}(\mathfrak{R},G) \longrightarrow \mathfrak{H}^q_v(\mathfrak{R},G)$ (II, 2.3). As a consequence of (21.1b) every Vietoris cycle $\delta^q \sim \tau\gamma^q$, where γ^q is a VS-cycle of $\tilde{\Sigma}^{p+1}$. Hence τ_1 is a homomorphism onto. We will show that it is also univalent. Since Σ^p_2 is essential in Σ^p a given

Σ^q_{vs} contains a cycle $\gamma^q \subset \Sigma^p_2$. Then $\nu\gamma^q = \gamma^q_1 \sim \gamma^q$, since ν is a metric singular deformation. Thus Γ^q_s contains a simplicial cycle which is in fact in $\tilde{\Sigma}^{p+1}$. If $\tau_1 \Gamma^q = 0$, then $\tau\gamma^q_1 = \mu\gamma^q_1 \sim 0$ in Ω^{p+1}_1 and since μ is a metric isomorphism $\gamma^q_1 \sim 0$ in $\tilde{\Sigma}^{p+1}$, hence also ~ 0 in $\mathfrak{R}$, and finally $\Gamma^q_{vs} = 0$. Thus τ_1 is univalent. Since it is a homorphism onto, it is an isomorphism and (22.2), hence (22.1) is proved.

(22.3) Let $\gamma^q = \{\gamma^q_n\}$ be a VS-cycle and Γ^q_{vs} its class. Then (II, 7.3) the γ^q_n are in a fixed Γ^q_s and $\Gamma^q_{vs} \longrightarrow \Gamma^q_s$ defines an isomorphism $\mathfrak{H}^q_{vs}(\mathfrak{R},G) \longrightarrow \mathfrak{H}^q_s(\mathfrak{R},G)$. If Γ^q_v is the Vietoris class adherent to Γ^q_{vs} in the sense of (II, 2.3) we may also agree to call Γ^q_v and Γ^q_s <u>adherent</u>. It will then be seen that we have proved more precisely:

> (22.4) When $\mathfrak{R}$ is an LC^p [LC^*] compactum then class-adherence for dimensions $\leq p$ [for all dimensions] determines an isomorphism between the corresponding singular and Vietoris groups.

23. The preceding results yield the following theorems due to Borsuk [c] and the author [d]:

> (23.1) THEOREM. When $\mathfrak{R}$ is an LC^p compactum then: (a) its discrete qth homology groups for $q < p - 1$, are subgroups of the corresponding groups of a certain finite simplicial complex Ψ. When $\mathfrak{R}$ is an LC^{pp} compactum all the groups for dimensions $\leq p$ are those of a point.

> (23.2) THEOREM. When $\mathfrak{R}$ is LC^* its discrete homology groups are subgroups of those of a certain finite simplicial complex Ψ. An

LC^{**} compactum is zero-cyclic.

(23.3) COROLLARY. The following statements may be made regarding the Betti-numbers $R^q(\mathfrak{R},\pi)$ (ordinary or mod π):

$\mathfrak{R}$ is LC^p : R^q is finite for $q \leq p$;

$\mathfrak{R}$ is LC^* : R^q is always finite and vanishes for q above a certain value;

$\mathfrak{R}$ is LC^{pp}: $R^0 = 1$, $R^q = 0$ for $1 \leq q \leq p$;

$\mathfrak{R}$ is LC^{**}: $R^0 = 1$, $R^q = 0$ for $q > 0$.

Let again the notations be those of (21.1). We have seen that every class Γ^q_s of finite singular cycles contains a γ^q of arbitrarily small mesh. Furthermore if γ^q of mesh $< \epsilon$ bounds it bounds a chain of diameter $< \epsilon$ (II, 5.4). Therefore the groups of (23.1) are the same, for $q \leq p - 1$, as those of Σ^p_2, where the mesh of Σ^p_2 is arbitrarily small. Suppose now $\mathfrak{R}$ to be LC^p. By (21.7) Σ^p_2 is singularly chain-deformable into the Ψ of (9.1), and so every class Γ^q_s contains a cycle γ^q of Ψ. Moreover the singular chain-deformation induces a homomorphism $\theta : \mathfrak{H}^q_s(\mathfrak{R},G) \longrightarrow \mathfrak{H}^q(\Psi,G)$. If $\theta\,\Gamma^q_s = 0$, i.e. if $\gamma^q \sim 0$ in Ψ then $\gamma^q \sim 0$ in $\mathfrak{R}$, and hence $\Gamma^q_s = 0$. Therefore θ is univalent and so it is an isomorphism of $\mathfrak{H}^q_s(\mathfrak{R}, G)$ with $\mathfrak{H}^q(\Psi, G)$. Since Ψ is simplicial this proves (23.1) for LC^p and $q \leq p - 1$.

Suppose now $q = p$. As before Γ^p_s is shown to contain a cycle γ^p of Ψ. It is possible however that $\gamma^p \sim 0$ in Ψ and yet $\Gamma^p_s \neq 0$, i.e. that θ is not a univalent homomorphism. Since every p-cycle of Ψ is in a $\theta\,\Gamma^p_s$, θ is still a homomorphism onto. Therefore $\mathfrak{H}^p(\mathfrak{R}, G)$ is isomorphic with the factor-group of $\mathfrak{H}^p(\Psi, G)$ by the kernel of θ. This completes the proof of (23.1) for LC^p.

When $\mathfrak{R}$ is LC^{pp} the complex Ψ may be chosen a point, proving (23.1) for LC^{pp}.

When $\mathfrak{R}$ is LC* the same Ψ may be chosen for all q and the restriction on q may be omitted. The treatment just given for LC^p and $q \leq p - 1$, applies then to all dimensions. This proves (23.2) for LC*. Finally when $\mathfrak{R}$ is LC** we may again take Ψ a point and omit the dimensional restrictions thus proving (23.2) for LC**.

(23.4) Since bounding in $\mathfrak{R}$ for a finite singular cycle is the same as bounding in Ψ and since Ψ is a sub-complex of Σ, we have by (AT,III,40.8):

> (23.5) If $\mathfrak{R}$ is an LC^p [LC*] compactum then the homology groups of $\mathfrak{R}$ for dimensions $\leq q$ [for all dimensions] are the corresponding groups of a certain finite complex Ψ with respect to bounding in another finite complex Ψ_1 of which Ψ is a closed subcomplex.

Referring to (AT,III,23.12), we find then that (AT, III,18.1) yields here:

> (23.6) If $\mathfrak{R}$ is an LC^p [LC*] compactum then the group of the integers is a universal group of coefficients for the homology groups for the dimensions $q \leq p$ [for all dimensions]

Naturally in the LC* case the groups for dimension $>$ dim Ψ are zero, so that in all cases at most a finite number of dimensions come under consideration.

24. <u>Groups at and around the points</u>. The same type of argument with minor modifications which leads to (22.1) will also yield:

> (24.1) When the compactum $\mathfrak{R}$ is LC^p

[LC^{ω} and *a fortiori* LC*] the corresponding Čech, Vietoris and singular homology groups for the cycles around a point and through a point are the same.

We also prove a theorem due in substance to P.S. Alexandroff [e] and Cech [c]:

(24.2) When $\mathfrak{R}$ is an LC^p compactum its homology groups for the cycles around the points and dimensions $q \leq p$ are those of a point. Hence an LC^{ω} compactum, and *a fortiori* an LC* compactum is zero-cyclic in all points (relative to the cycles around the points).

As a consequence we have the following relations for the Betti-numbers around the points:

(24.3) $\mathfrak{R}$ is LC^p : $R^0(x,\pi) = 1$, $R^q(x,\pi) = 0$, $0 < q \leq p$;

(24.4) $\mathfrak{R}$ is LC : $R^0(x,\pi) = 1$; $R^q(x,\pi) = 0$, $q > 0$.

Let $\mathfrak{R}$ be LC^p and let us return for a moment to the situation and notations of (9). Take first a finite open $\zeta/2$ covering $\mathfrak{B} = \{V_i\}$ and modify it as follows: all the sets $V_i \not\ni x$ remain the same; those $\ni x$ are united into a single open set. The resulting new covering $\mathfrak{W}$ will be of mesh $< \zeta$ and will have the property that x is contained in one and only one set of $\mathfrak{W}$. If we now take an irreducible refinement $\mathfrak{U}$ of $\mathfrak{W}$ such that nerve $\overline{\mathfrak{U}}$ = nerve $\mathfrak{U}$, it will be found by reference to (AT,VII, proof of 3.3) that x will still be in a single set of $\mathfrak{U}$, say U_1. Therefore referring to (9) we may assume that x is the vertex A_1 of Φ. We will then have an $\mathfrak{S}(x,\rho) \subset U_1$ and such that $\mathfrak{S}(x', \rho) \cap U_i = \emptyset$, $i > 1$. It follows that $s\,\mathfrak{S}(x,\rho) = A_1 = x$ (s as in 9). By reference to (9) we find then that if $\gamma^q = \{\gamma_n^q\}$,

$q \leqq p$, is a VS-cycle then for n above a certain value γ_n^q will be homotopic with a cycle $\subset x$ say in $\mathfrak{S}(x, 2\epsilon)$. Hence $\gamma^q \sim 0$. This proves (24.2) for LC^p. When $\mathcal{R}$ is LC^ω the same argument holds for every p, i.e. no restrictions need be placed on q. This completes the proof of (24.2).

§6. COINCIDENCES AND FIXED POINTS

25. These questions have already been dealt with in (AT,VIII,29) for finite polyhedra, and (AT,VIII,34) for a general class of spaces, the so-called quasi-complexes. The LC^* type to be considered exclusively here, may be shown to consist of quasi-complexes. However rather than rest the case upon the results of (AT, VIII,34) it is more expedient to give a direct proof utilizing singular cycles.

Consider first a mapping t of $\mathcal{R}$ into itself. Referring to (23.1) the rational homology groups of $\mathcal{R}$ for dimensions exceeding a certain n are zero, and those for dimension $\leqq n$ have finite bases. Let $\{\gamma_i^q\}$, $i = 1, 2, \ldots, R^q$ be a base for the rational q-cycles, $q \leqq n$. Since t induces a homomorphism of each rational group $\mathfrak{H}^q$ into itself, we have homologies with rational coefficients:

$$(25.1) \qquad t\,\gamma_i^q \sim \lambda_i^j(q)\,\gamma_j^q, \quad 0 \leqq q \leqq n,$$

and the number

$$(25.2) \qquad \psi(t) = \sum (-1)^q \ \text{trace} \ \lambda(q), \quad \lambda(q) = \|\lambda_i^j(q)\|$$

is well defined and independent of the particular bases chosen. We have in fact from (AT,VIII,28.9):

(25.3) The numbers trace $\lambda(q)$ $\psi(t)$ depend solely upon the homotopy class of t.

We shall now prove:

(25.4) **FIXED POINT THEOREM.** Let $\mathfrak{R}$ be an LC* compactum and t any mapping of $\mathfrak{R}$ into itself. If $\psi(t) \neq 0$ then t has a fixed point (Lefschetz [e]).

Let Ψ be the continuous complex of (9.1) and (23.1) corresponding to a certain $\epsilon > 0$. By (23.1) we may assume that the γ_i^q are cycles of Ψ itself. To have a base for the rational q-cycles of Ψ it may be necessary to add a certain set of ρ^q cycles $\{\gamma'^q_i\}$. However we have

$$\gamma'^q_i \sim a_i^j \; \gamma_j^q \quad \text{in } \mathfrak{R}.$$

Since we may manifestly replace γ'^q_i by $\gamma'^q_i - a_i^j \gamma_j^q$ we may assume the γ'^q_i so chosen that $\gamma'^q_i \sim 0$ in $\mathfrak{R}$. We have then a set of singular chains C_i^{q+1} such that $FC_i^{q+1} = \gamma'^q_i$. Consider now the singular complex $\mathfrak{L}$ which is the union of Ψ and all the $|C_i^{q+1}|$ (for all q) and its transform $t\mathfrak{L}$. Since $\mathfrak{L}$ is finite by (23.1) there is an r such that $(tL)^{(r)}$ is ϵ singularly chain-deformable into Ψ. If τ is the singular chain-deformation then t. and $\tau t = s$ map any cycle of Ψ into cycles homologous in $\mathfrak{R}$. Hence

(25.5) $s\,\gamma_i^q \sim \lambda_i^j\,(q)\,\gamma_j^q + \mu_i^j\,(q)\,\gamma'^{\,q}_j$ in Ψ.

On the other hand $s\,\gamma'^q_i = s\,FC_i^{q+1} = F(s\,C_i^{q+1})$ and hence

(25.6) $$s\,\gamma'^q_i \sim 0 \quad \text{in } \Psi.$$

Referring to (9) we have found $\Psi = (\Phi, T)$, where Φ is an Euclidean nerve of the covering $\mathfrak{U}$ of (9). Let $\Phi = \{\sigma\}$ and $E = (\sigma, T)$. As constructed we have found mesh $\Psi < \eta(\frac{1}{3}\epsilon)$, and so

(25.7) $$\operatorname{diam} \|E\| < \eta(\tfrac{1}{3}\epsilon) < \epsilon.$$

On the other hand since s is an ϵ displacement we have

(25.8) $$d(\|E\|, \|s\,E\|) \geq d(\|E\|, \|tE\|) - \epsilon.$$

These inequalities will be needed in a moment. We first prove:

(25.9) If t has no fixed point there is a $\zeta > 0$ such that every $d(x,tx) > \zeta$.

For given x let $2\rho = d(x,tx) > 0$. Since t is continuous there is a $\nu > 0$ such that $t\,\mathfrak{S}(x,\nu) \subset \mathfrak{S}(x,\rho)$, and so $\mathfrak{S}(x,\nu) \cap t\,\mathfrak{S}(x,\nu) = \emptyset$. Since $\mathfrak{R}$ is a compactum the open covering $\{\mathfrak{S}(x,\nu)\}$ has a finite subcovering $\{S_i\}$ and we have $S_i \cap tS_i = \emptyset$ throughout. Let 2ζ be the Lebesgue number of $\{S_i\}$. Then whatever x the sphere $\mathfrak{S}(x,\zeta)$ is in an S_i and so $\mathfrak{S}(x,\zeta) \cap t\,\mathfrak{S}(x,\zeta) = \emptyset$, hence $d(x,tx) > \zeta$.

Returning to our problem suppose t without fixed points. Since $\mathfrak{R}$ is a compactum t is uniformly continuous. Hence we may choose $\epsilon < \zeta/4$ such that $\operatorname{diam} A < \epsilon \Longrightarrow \operatorname{diam} tA < \zeta/4$. As a consequence from (25.8): $d(\|E\|, \|sE\|) \geq \zeta/4$. Hence E may never be a fixed element of s. Therefore if we designate by ψ_1 the ψ_1 of (AT,V,§3), we must have $\psi_1(s) = 0$. From (25.5, 25.6) and (AT,V, 24.12b) there follows however $\psi_1(s) = \psi(t)$. Hence $\psi_1(s) = 0 \Longrightarrow \psi(t) = 0$. Thus when t has no fixed points $\psi(t) = 0$, and this proves (25.4).

An incidental result is:

(25.10) The number $\psi(t)$ is an integer or zero.

For $\psi(t) = \psi_1(s)$ and $\psi_1(s)$ is an integer or zero (AT,V,22.11).

26. Let now $\mathfrak{R}, \mathfrak{S}$ be compacta one of which say $\mathfrak{R}$ is LC*. Let us assume that whenever the Betti numbers of $\mathfrak{R}$ are $\neq 0$ those of $\mathfrak{S}$ are finite. We have an integer n such that $R^{n+1}(\mathfrak{R}) = 0$, and that $R^{n-j}(\mathfrak{R}), j \leq n$,

is finite. Thus for $q \leqq n$, both $\mathfrak{R}$ and $\mathfrak{S}$ have finite bases $\{\gamma_i^q\}$, $\{\gamma_i^q\}$ for the rational (Vietoris or singular) q-cycles, and every cycle of $\mathfrak{R}$, whose dimension $> n$, ~ 0.

Suppose now that we have two mappings $t\colon \mathfrak{R} \to \mathfrak{S}$, $t_1 : \mathfrak{S} \longrightarrow \mathfrak{R}$. We have then homologies for $q \leqq n$:

$$t\,\gamma_i^q \sim \lambda_i^j\,(q)\,\delta_j^q,\quad \lambda(q) = \|\lambda_i^j(q)\|$$

$$t_1\delta_i^q \sim \mu_i^j\,(q)\,\gamma_j^q,\quad \mu(q) = \|\mu_i^j(q)\|\ .$$

Hence

$$t_1 t\,\gamma_i^q \sim \lambda_i^j\,(q)\ \mu_j^k\,(q)\,\gamma_k^q.$$

Let us set

(26.1) $\phi(t,\ t_1) = \psi(t_1 t) = \sum (-1)^q$ trace $\lambda(q)\,\mu(q)$.

Here again by (AT,VIII,28.9):

> (26.2) The numbers trace $\lambda(q)\,\mu(q)$, and hence also $\phi(t,\ t_1)$ depend solely upon the homotopy classes of t and t_1.

By an observation already made t and t_1 have a coincidence (x,y), $x \in \mathfrak{R}$, $y \in \mathfrak{S}$, when and only when $t_1 t$ has the fixed point x. It follows that when $\phi(t,\ t_1) \neq 0$ then $\psi(t_1, t) \neq 0$ and hence t and t_1 have a coincidence. This proves:

> (26.3) THEOREM. Let $\mathfrak{R}_1\ \mathfrak{S}$ be compacta, where $\mathfrak{R}$ is LC* and $\mathfrak{S}$ has the property that for the values q such that the rational Betti number $R^q(\mathfrak{R}) \neq 0$ then $R^q(\mathfrak{S})$ is finite, a condition fulfilled when $\mathfrak{S}$ is likewise LC*. If $t,\ t$ are mappings $\mathfrak{R} \longrightarrow \mathfrak{S}$, $\mathfrak{S} \longrightarrow \mathfrak{R}$ such that $\phi(t,\ t_1) \neq 0$, then t and t_1 have a coincidence.

27. <u>Applications</u>. We have already given a number of applications in (AT,VIII,30). Before considering

others we need the

(27.1) DEFINITION. A topological space $\mathfrak{R}$ is said to have the <u>fixed point property</u> whenever every mapping of $\mathfrak{R}$ into itself has a fixed point.

(27.2) If one of two spaces $\mathfrak{R}, \mathfrak{S}$ has the fixed point property then any two mappings $t : \mathfrak{R} \longrightarrow \mathfrak{S}$ and $t_1 : \mathfrak{S} \longrightarrow \mathfrak{R}$ have a coincidence.

For if say $\mathfrak{R}$ has the fixed point property then $t_1 t : \mathfrak{R} \longrightarrow \mathfrak{R}$ has a fixed point x, and hence t and t_1 have a coincidence (x, tx).

(27.3) The following compacta have the fixed point property: (a) absolute retracts; (b) connected absolute neighborhood retracts whose Betti-numbers R^q, $q > 0$, vanish.

For a space $\mathfrak{R}$ of one or the other type is zero-cyclic. Every point x defines then a Vietoris zero-cycle $\gamma^0 = x$. Furthermore every discrete Vietoris $\gamma^0 \sim \lambda x$. In particular, if t is a mapping $\mathfrak{R} \longrightarrow \mathfrak{R}$ then $tx \sim \lambda x$. Now if $\gamma^0 = \{\gamma^0_n\}$, $\gamma^0_n = x$, then $t\gamma^0 = \{\delta^0_n\}$, $\delta^0_n = x'$ where $x' = tx$. Therefore the cycle $x' \sim x$, or $tx \sim x$, $\lambda = 1$. As a consequence $\psi(t) = 1 \neq 0$, and t has a fixed point (25.4).

(27.4) If one of the spaces $\mathfrak{R}, \mathfrak{S}$ is of one of the two types of (27.3), then any two mappings $t: \mathfrak{R} \longrightarrow \mathfrak{S}$, $t_1 : \mathfrak{S} \longrightarrow \mathfrak{R}$ have a coincidence (27.2, 27.3).

(27.5) Every closed convex subset of the

Hilbert parallelotope P^{ω} has the fixed point property (10.1, 12.2b, 27.3).

28. Applications to functional analysis. The fixed point property may be utilized to advantage to obtain analytical existence theorems. This was first done by Birkhoff-Kellogg. (Transactions Am. Math. Soc., 23 (1922), 96 — 115). The strongest results of this nature obtained so far are due to Schauder [a]. We shall discuss his three basic theorems and show that they reduce to (27.3a), hence to the author's basic theorem (25.4).

The spaces under consideration are subsets of metric vector spaces of the type searchingly investigated by Banach to whom we owe the standard work on the subject: Opérations Linéaires (Warsaw, 1932). We shall first recall a few concepts needed here.

Let then $\mathfrak{R}$ be a real vector space in the sense of (AT,II,22.1) whose topology is given by a metric $d(x,y)$. We call $\mathfrak{R}$ a space of type Φ. It may be observed that those spaces of type Φ which are complete and satisfy

(28.1) $$d(x,y) = d(x-y,0)$$

are the spaces of type F of Banach (loc. cit. Chapter III), as shown in a paper by Mazur and Orlicz (Studia Mat., 4 (1933), 152-158; see p. 154). A norm of $\mathfrak{R}$, is a real valued function $\|x\|$ such that

(28.2)
$$\begin{aligned} &(a)\ \|x\| \geqq 0;\ \|x\| = 0 \Longrightarrow x = 0; \\ &(b)\ \|kx\| = |k| \cdot \|x\|; \\ &(c)\ \|x + y\| \leqq \|x\| + \|y\| . \end{aligned}$$

A vector space with a given norm is said to be normed. It is immediately apparent that when $\mathfrak{R}$ has a norm x then $d(x,y) = \|x - y\|$ is a metric for $\mathfrak{R}$ under which it is of type Φ. If also $\mathfrak{R}$ is complete under this metric it is known as a Banach space (type B of Banach's own nomenclature).

Let $\mathfrak{R}$ be a Banach space. A <u>linear functional</u> f over $\mathfrak{R}$ is essentially a continuous field-character in the sense of (AT,II,28.1). That is to say f is a continuous and real-valued function defined over $\mathfrak{R}$, satisfying

$$f(\alpha x + \beta y) = \alpha f(x) + \beta f(y) \quad (x, y \in \mathfrak{R} : \alpha, \beta \text{ real}).$$

It can be shown that the set of all linear functionals over $\mathfrak{R}$ is also a Banach space, $\mathfrak{R}^*$, the <u>conjugate</u> of $\mathfrak{R}$. The <u>norm</u> in $\mathfrak{R}^*$ is taken to be $\|f\| = \sup |f(x)|$ for $\|x\| \leq 1$, and it is necessarily finite.

Since $\mathfrak{R}$ is normed, it has a topology as a metric space (of type Φ). This is sometimes called the "<u>strong</u>"topology to distinguish it from a topology (different, in general) called the "<u>weak</u>" topology, which is defined by means of $\mathfrak{R}^*$ as follows: If $\{l\}$ are the intervals of the real line, and $\mathfrak{R}^* = \{f\}$, then $\{f^{-1}l\}$ is a sub-base for the weak topology of $\mathfrak{R}$. The prefix "weak" is used to indicate that this topology is implied. Thus a set $A \subset \mathfrak{R}$ is "weakly closed" if it is closed in the weak topology, etc.

It is easily verified that a necessary and sufficient condition that, in $\mathfrak{R}$, $\{x_n\}$ be weakly convergent to x, is that $f(x_n) \longrightarrow f(x)$ for each $f \in \mathfrak{R}^*$.

Similarly the topology in $\mathfrak{R}^*$ as a normed vector space is sometimes called its "strong" topology, in contrast with its "weak" topology which is defined in terms of $\mathfrak{R}$ as follows: If $\{l\}$ are the intervals of the real line, and $\mathfrak{R} = \{x\}$, then $\{f \mid f(x) \in l\}$ is a sub-base for the weak topology in $\mathfrak{R}^*$.

As before, a necessary and sufficient condition that, in $\mathfrak{R}^*$, $\{f_n\}$ be weakly convergent to f, is that $f_n(x) \longrightarrow f(x)$ for each $x \in \mathfrak{R}$.

We shall also need the

(28.3) DEFINITION. The set $A \subset \mathfrak{R}$ is said to be sequentially <u>conditionally</u> compact whenever every sequence $\{x_n\} \subset A$ has a subsequence $\{x_{n'}\}$ convergent to a point x of $\mathcal{R}$.

29. We may now state Schauder's theorems.

(29.1) Every convex compactum H in a space of type Φ has the fixed point property.

(29.2) If H is convex and closed in a Banach space and t is a mapping $H \longrightarrow H$ such that $H_1 = tH$ is conditionally compact, then t has a fixed point.

(29.3) Let $\mathcal{R}$ be a strongly separable Banach space and H a strongly closed and convex subset of $\mathcal{R}$ which is weakly sequentially conditionally compact. Then every weakly continuous mapping $t : H \longrightarrow H$ has a fixed point.

Theorem (29.3) has been generalised by Krein and Smulian to:

(29.4) If H is a strongly closed convex subset of a Banach space, and t is a weakly continuous mapping of H into H such that $H_1 = tH$ is separable and weakly sequentially conditionally compact, then t has a fixed point.

(29.5) Remark. Schauder's first proposition (29.1) includes the classical theorems of Birkhoff-Kellogg already referred to in (28).

Proof of (29.1). We will say that the real vector space $\mathcal{R}$ with a metric is radially locally contractible whenever given any $\epsilon > 0$ and $x_0 \in \mathfrak{R}$ there exists an $\eta > 0$ such that $\mathfrak{S}(x_0, \eta)$ is deformable into x_0 over $\mathfrak{S}(x_0, \epsilon)$ and so that each point x has for path the segment $\overline{xx_0}$.

(29.6) The space $\mathfrak{R}$ of (29.1) is radially locally contractible.

Given any $\eta > 0$ let $\zeta(\eta) = \sup \{d(x_0, (1 - k)x_0 + kx) \mid x \in \mathfrak{S}(x_0, \eta), 0 \leq k \leq 1\}$. It is clear that $\zeta(\eta)$ is a non-increasing function. We will show that if $\{\eta_n\} \rightarrow 0$ then $\{\zeta(\eta_n)\} \rightarrow 0$ also. For in the contrary case $\inf \zeta = \zeta_0 > 0$ and so for each n there is an $x_n \in \mathfrak{S}(x_0, \frac{1}{n})$ and a $k_n \in [0 - 1]$ such that $d(x_0, (1 - k_n)x_0 + k_n x_n) \geq \cdot \zeta_0$. We choose a subsequence $\{n'\}$ of $\{n\}$ such that $\{k_{n'}\}$ converges say to k_0. Since $\{x_{n'}\} \longrightarrow x_0$ and kx is continuous, there is a ν such that $n' > \nu \Longrightarrow d(x_0, (1 - k_{n'})x_0 + k_{n'} x_{n'}) < \frac{1}{2} \zeta_0$ and this contradiction proves (29.6).

(29.7) A convex compactum H in a space of type Φ is contractible. (Same proof as in AT, I, 47.4).

(29.8) The set H of (29.7) is locally contractible.

For let $x_0 \in H$. Given any $\epsilon > 0$ there is an $\eta > 0$ such that $\mathfrak{S}(x_0, \eta)$ is rectilinearly contractible to x_0 in $\mathfrak{S}(x_0, \epsilon)$. Since the paths are segments $\mathfrak{S}(x_0, \eta) \cap H$ is contractible to x_0 in $\mathfrak{S}(x_0, \epsilon) \cap H$ which proves (29.8).

Since H of (29.1) is contractible and locally contractible it is an AR and so (29.1) is a consequence of (27.3a).

Proof of (29.2). The convex closure $\widehat{H}_1$ of the conditionally compact subset H_1 (of a Banach space) is compact (Mazur, Studia Mathematica, vol. 2 (1930, pp. 7-9). Since $\overline{H}_1 \subset H$, $\widehat{H}_1$ is a convex compactum mapped into itself by t. Therefore by (29.1), there is an $x \in \widehat{H}_1$, hence $\in H$, such that $tx = x$.

Proof of (29.3). We shall show, (following Tukey) that under the conditions considered H may be assigned a norm giving rise to a metric which induces the weak convergence in H. According to Mazur (Studia Mat., 4 ('933), pp. 70-84, Theorem 3), H is weakly closed and hence weakly sequentially compact. Since $\mathfrak{R}$ is strongly separable $\mathfrak{R}^*$ is weakly sequentially separable (Banach: loc. cit., p. 124, Theorem 4). That is to say there exists a sequence of functionals $\{f_n\}$ such that for every $f \in \mathfrak{R}^*$ and some $\{n_k\}$ we have $\{f_{n_k}(x)\} \longrightarrow f(x)$ for every $x \in \mathfrak{R}$.

Also it can be supposed that each $f_n \neq 0$. Suppose now $x \in \mathfrak{R}$ and $\{x_m\} \subset H$ such that for every n we have $\lim f_n(x_m) = f_n(x)$. We prove that $x_m \longrightarrow x$ (weakly). If not, there is a linear functional F such that $F(x_m) \not\rightarrow F(x)$, and so there is a subsequence $\{x_{m'}\}$ for which $F(x_{m'}) \longrightarrow A \neq F(x)$ (where A is finite or infinite). Since H is weakly sequentially (conditionally) compact, there is a subsequence $\{x_{m''}\}$ of $\{x_{m'}\}$ and an $x' \in \mathfrak{R}$, for which $x_{m''} \longrightarrow x'$ weakly. Hence, for each n, $f_n(x_{m''}) \longrightarrow f_n(x')$. So $f_n(x) = f_n(x')$ for each n. Since the f_n's are weakly dense in R^*, it follows that $F(x) = F(x')$. And we have $A = \lim F(x_{m'}) = \lim F(x_{m''}) = F(x') = F(x)$, which is a contradiction. This proves that $x_m \longrightarrow x$ (weakly).

Let now $\|x\|_1 = \sum \dfrac{|f_n(x)|}{2^n \|f_n\|}$, and $d_1(x,y) = \|x - y\|_1$. (The series for $\|x\|_1$ necessarily converges.) It is clear that $\|x\|$ is a norm for H. By what has just been shown, if $\{x_n\} \in H$ and $d_1(x, x_n) \longrightarrow 0$, then

$f(x_n) \longrightarrow f(x)$ for all $f \in \mathfrak{R}^*$. The converse is obvious. Therefore $d_1(x,y)$ induces the weak convergence in H. Since H is weakly sequentially compact, it follows that H with the metric d_1 is a sequentially compact metric space, and hence is a compactum. Also t is manifestly continuous with respact to this metric. Since H is convex and its metric (d_1) is a norm, H is manifestly locally radially contractible and also contractible (29.7, 29.8). Therefore (29.3) is again a consequence of (27.3a).

Proof of (29.4). Let H be a (strongly) closed common subset of a Banach space, mapped by t (a weakly continuous mapping) onto $tH = H_1 \subset H$. Consider the convex closure $\widehat{H}_1$ of H_1. If H_1 is weakly sequentially conditionally compact, so is $\widehat{H}_1$. (For a proof, see Krein-Šmulian, Annals of Math., vol. 41 (1940 pp. 556-583, esp. p. 582). Let R_1 be the closed linear manifold determined by H_1. If H_1 is separable, it is easily verified that R_1 is separable. Also, clearly $\widehat{H}_1 \subset H$. So the theorem follows on applying (29.3) to the mapping t of $\widehat{H}_1$ as a subset of R_1.

§7. HLC SPACES. GENERALIZED MANIFOLDS

30. In this concluding section we shall briefly recall certain extensions of the LC concepts which have been discussed more fully elsewhere notably by Alexandroff, Čech, Wilder and the author. A thorough treatment by E. G. Begle [a] is to appear shortly containing many new results, and in particular a very full discussion of generalized manifolds, and it is to this paper that the reader is referred for all details.

31. A natural extension of the first LC concepts suggested by the homotopy-homology analogies is obtained if spheres are replaced by cycles and contractibility by "~ 0". It is evident however from (§2), that the method

of partial realization is the more powerful and the analogy here is to replace continuous complexes by images of complexes under chain-mappings. We have thus a great deal of latitude according to the types of chain-mappings which may be considered, etc. To simplify matters we consider only compact spaces and the concepts related to the Čech homology theory.

Let then $\mathfrak{R}$ be compact and let $\{\mathfrak{U}_\lambda\}$, $\{\Phi_\lambda\}$ be the finite open coverings of $\mathfrak{R}$ and their nerves. For convenience we denote by $\mathfrak{U}_{\lambda+\mu}$ the open covering $\mathfrak{U}_\lambda \cup \mathfrak{U}_\mu$. Let also the homology theory of $\mathfrak{R}$ and all the complexes considered be taken augmented (AT,VII,3). Consider now a generalized chain-mapping τ over a given group $G: K \longrightarrow \Phi_\lambda$ (see AT,IV,13). We call (K,τ) a λ-realization of K over G. The chain $\tau\sigma$ consists of a certain number of simplexes of Φ_λ. If it is true that for for some μ and all σ each chain $\tau\sigma$ is contained in a set of $\mathfrak{U}_\mu$ (see AT,VII,4) we say that the mesh of the realization (K,τ) is $\prec \mu$.

Suppose now that L is a dense closed subcomplex of K and let τ be a generalized chain-mapping $L \longrightarrow \Phi_\lambda$ over G. We call (L,τ) a partial λ-realization of K over G. If $\sigma \in K$, then $L \cap \mathrm{Cl}\,\sigma$ is imaged into a collection of chains of Φ_λ. Whenever for some μ and all σ all the chains of this collection are contained in a set of $\mathfrak{U}_\mu$ we say that the mesh of the partial realization (D,τ) is $\prec \mu$. If there exists a realization (K,τ) such that $\tau_1 = \tau | L$ we say that it extends the given partial realization (L,τ).

32. We may now state the:

(32.1) DEFINITION. The compact space $\mathfrak{R}$ is said to be HLC^p over G whenever corresponding to every λ,μ there is a $\lambda'(\lambda,p)$ and a $\mu'(\lambda,\mu,p)$ such that every λ'-partial realization over G of a simplicial complex K whose dimension $\leqq p + 1$ and whose mesh $\prec \mu'$ may be extended to a $(\lambda+\lambda')$-realization whose mesh $\prec \mu$. If the preceding property holds with $\lambda'(\lambda)$ and $\mu'(\lambda,\mu)$ independent of p, then $\mathfrak{R}$ is said to be HLC*. If in addition all the homology groups over G of $\mathfrak{R}$ augmented are zero, we say that $\mathfrak{R}$ is HLC**. The analogy with (6.3, 6.4) is clear.

We state the following properties for which the reader is referred to Begle:

(32.2) If $\mathfrak{R}$ is HLC^p over G and $\dim \mathfrak{R} \leqq p$ then $\mathfrak{R}$ is HLC*.

(32.2') The homology groups over G of an HLC^p space over G(G discrete) are isomorphic with subgroups of those of a finite complex. If $\mathfrak{R}$ is HLC* over G this holds for all the groups over G. When G is a field the Betti-numbers of an HLC^p over G for dimensions $\leqq p$ are all finite, while for an HLC* over G this holds for all dimensions, those above a certain dimension being zero.

(32.3) If $\mathfrak{R}$ is HLC^p (HLC*) over the integers it is HLC^p (HLC*) over any discrete group whatever. (Čech).

(32.4) The fixed point theorem (25.4) is valid for a compact space which is HLC* over the rationals. Similarly the coincidence theorem (28.3) holds with LC* replaced by HLC*.

(32.5) Let G be a field and let $R^q(x,G)$ denote the Betti-number for the q-cycles over G around the point x. Then a n.a.s.c. for $\mathcal{R}$ to be HLC^p over G is that for all x: $R^q(x,G) = 0$ for $0 < q < p$, and $R^o(x,G) = 1$.

An immediate corollary is

(32.6) If $\mathcal{R}$ is HLC^p over a field G it is HLC^p over any other field with the same characteristic as G.

As a consequence if π is the characteristic of G instead of "HLC^p over G" we may say "HLC^p mod π".

The last two results do not "reach out" to the HLC* types. It so happens that in the extension to cocycles these types are unimportant, and furthermore that the field cases are those which chiefly matter. This suggests then the simple

(32.7) DEFINITION. Let $R_q(x,\pi)$ denote the qth Betti number for the cocycles around x mod π, where π is prime. Then the compact space $\mathcal{R}$ is said to be HLC_p whenever for all x: $R_q(x,\pi) = 0$ for $0 < q \leqq p$, and $R_o(x,\pi) = 1$.

We have now all the elements for the definition of generalized manifolds given by Begle:

(32.8) DEFINITION. The compact space $\mathcal{R}$ is said to be <u>an absolute generalized n-manifold mod π</u> whenever the following conditions subsist:

(a) $\dim \mathcal{R} = n$;

(b) $\mathcal{R}$ is HLC^n mod π;

(c) $\mathfrak{R}$ is HLC_{n-1} mod π ;

(d) $R_n(x,\pi) = 1$ for every x.

If $\pi \neq 2$ and $\mathfrak{R}$ is connected, we have $R^n(\mathfrak{R},\pi) = 0$ or 1. In the first case the manifold is said to be non-orientable, in the second orientable. If $\mathfrak{R}$ is not connected it is said to be orientable whenever all its components are orientable, and it is said to be non-orientable otherwise.

It has been proved by Begle that:

(32.9) When $\mathfrak{R}$ is an orientable absolute generalized n-manifold then $R^p(\mathfrak{R},\pi) = R^{n-p}(\mathfrak{R},\pi)$ (duality in the sense of Poincaré).

We also refer the reader to Begle's paper for two examples showing that: (a) the HLC properties are weaker than the LC properties; (b) there exist generalized manifolds of dimension $n \geq 3$ which are not topological manifolds (i.e. with points which have no n-cell for neighborhood).

SPECIAL BIBLIOGRAPHY ON LOCAL CONNECTEDNESS AND RETRACTION

(Starred articles contain material on retraction)

ALEXANDROFF, P.

1. On local properties of closed sets. Ann. of math. 36:1-35 (1935).
2. Zur Homologie-Theorie der Kompakten. Compositio math. 4:256-270 (1937).

ALEXANDROFF, P. and PONTRJAGIN, L.

1. Les variétés à n-dimensions généralisées. Acad. sci., Paris. Compt. rend. 202:1327-1329 (1936).

ARONSZAJN, N.

1. *Sur les lacunes d'un polyèdre et leurs relations avec les groupes de Betti. Akad. van wetens. Amsterdam, Proc. 40:61-69 (1937).

ARONSZAJN, N., et BORSUK, K.

1. *Sur la somme et le produit combinatoire des rétractes absolus. Fundam. math. 18:193-197 (1932).

AYERS, W. L.

1. *Some generalizations of the Scherrer fixed-point theorem. Fundam. math. 16:332-336 (1930).

BEGLE, E. G.

1. Locally connected spaces and generalized manifolds. Amer. jour. math. vol. 64:553-574 (1942).

BORSUK, K.

1. *Sur les rétractes. Fundam. math. 17:152-170 (1931).
2. Einige Sätze über stetige Streckenbilder. Fundam. math. 18:198-213 (1932).

3. Über eine Klasse von lokal zusammenhängenden Räumen, Fundam. math. 19:220-242 (1932).

4.*Zur kombinatorishen Eigenschaften der Retrakte. Fundam. math. 21:91-98 (1933).

5. Zur Dimensionstheorie der lokal zusammenziehbaren Räume, Math. ann. 109:376-380 (1934).

6. Sur un continu acyclique qui se laisse transformer en lui même sans points invariants. Fundam. math. 24:51-58 (1935).

7.*Quelques rétractes singuliers, Fundam. math. 24:249-258 (1935).

8. Un théorème sur les groupes de Betti des ensembles localement connexes en toutes les dimensions $\leq$ n. Fundam. math. 24:311-316 (1935).

9. Ensembles dont les dimensions modulaires de Alexandroff coïncident avec la dimension de Menger-Urysohn, Fundam. math. 27:77-93 (1936).

10.*Sur le plongement des espaces dans les rétractes absolus. Fundam. math. 27:239-243, (1936).

11. Sur les transformations continues n'augmentant pas la dimension. Fundam. math. 28. 90-98 (1937).

12. Sur les prolongements des transformations continues. Fundam. math. 28:99-110 (1937).

13. Über spheroïdale und H-spheroidale Räume. Matem. sbornik, n.s., 1:643-660 (1936).

14. Un théorème sur le prolongement des transformations, Fundam. math. 29:161-166 (1937).

15. Un theoreme sur le prolongement des fonctions continues. Polskie towarz. mat., Annales 16:218 (1937).

16.*Quelques relations entre la situation des ensembles et la rétraction dans les espaces euclidiens. Fundam. math. 29:191-205 (1937).

17. Sur les coupures locales des variétés. Fundam. math 32:288-293 (1939).

BORSUK, K., et MAZURKIEWICZ, S.

1. Sur les rétractes absolus indécomposables. Acad. sci., Paris, Compt. rend. 199:110-112 (1934).

CECH, E.

1. Sur les nombres de Betti locaux. Ann. of math. 35:678-701 (1934).
2. Sur la connexité locale d'ordre supérieur. Compositio math. 2:1-25 (1935).

EILENBERG, S.

1.*Sur les courbes sans noeuds. Fundam. math. 28: 233-242 (1937).
2.*Un théorème sur l'homotopie. Ann. of math. 38: 656-661 (1937).

FOX, R. H.

1. On homotopy and extension of mappings. Nat. acad. sci. Proc. 26:26-28 (1940).
2.*A characterization of absolute neighborhood retracts. Amer. math. soc. Bull. 48:271-275 (1942).

HUREWICZ, W.

1. Homotopie, Homologie und lokaler Zusammenhang. Fundam. math. 25:467-485 (1935).
2. Beiträge zur Topologie der Deformationen:
   I. Höherdimensionale Homotopiegruppen. Akad. wetensch. Amsterdam, Proc. 38:112-119 (1935).
   II. Homotopie-und Homologiegruppen. Ibid. 38:521-528 (1935).
   III. Klassen und Homologietypen von Abbildungen. Ibid. 39:117-126 (1936).
   IV. Asphärische Räume. Ibid. 39:215-224 (1936).
3. Homotopie und Homologie. Matem. sbornik, n. s., 1: 697 (1936).

KELLEY, J. L.

1. Hyperspaces of a continuum. Amer. math. soc. Trans. (to appear soon).

KURATOWSKI, K.

1. Sur les espaces localement connexes et peaniens en dimension n. Fundam. math. 24:269-287 (1935).
2. La notion de connexité locale en topologie. Enseignement math. 35:229-240 (1936).
3. Une condition métrique pour le rétraction des ensembles. Towarz. nauk. Warszaw. Cl. III Compt. rend. 28:156-158 (1936).
4. Sur la compactification des espaces à connexité n-dimensionelle. Fundam. math. 30:242-246 (1938).

KURATOWSKI, K., et OTTO, E.

1. Sur les espaces à connexité n-dimensionelle. Fundam. math. 32:259-264 (1939).

LEFSCHETZ, S.

1. On generalized manifolds. Amer. jour. math. 55: 469-504 (1933).
2. Chain deformations in topology. Duke math. jour. 1: 1-18 (1935).
3. *Locally connected and related sets. I. Ann. of math. 35:118-129 (1934).
4. *Locally connected and related sets. II. Duke math. jour. 2:435-442 (1936).
5. Locally connected sets and their applications. Matem. sbornik, n.s., 1:715-717 (1936).
6. On the fixed point formula. Ann. of math. 38:819-822 (1937).
7. *Locally connected sets and retracts. Nat. acad. sci., Proc. 24:392-393 (1938).
8. Singular and continuous complexes, chains and cycles. Matem. sbornik, n.s., 3:271-285 (1938).
9. Chains of topological spaces. Ann. of math. 39: 383-396 (1938).

PUCKETT, W. T. JR.

1. Regular transformations. Duke math. jour. 6:80-88 (1940).

VAUGHN, H. E. JR.

1. On local Betti numbers. Duke math. jour. 2:117-137 (1936).

WHYBURN, G. T.

1. Cyclic elements of higher orders. Amer. jour. math. 56:133-146 (1934).
2. Regular convergence and monotone transformations. Amer. jour. math. 57:902-906 (1935).
3. Sequences and limiting sets. Fundam. math. 25: 408-426 (1935).
4. *Non-alternating interior retracting transformations. Ann. of math. 40:914-921 (1939).

WILDER, R. L.

1. Point sets in three and higher dimensions and their investigation by means of a unified analysis situs. Amer. math. soc., Bull. 38:649-692 (1932).
2. Concerning a problem of K. Borsuk. Fundam. math. 21:156-167 (1933)
3. On the properties of domains and their boundaries in E_n. Math. ann. 109:273-306 (1933).
4. Generalized closed manifolds in n-space. Ann. of math. 35:876-903 (1934).
5. Locally connected spaces. Duke math. jour. 1:543-555 (1935).
6. On free subsets of E_n. Fundam. math. 25:200-208 (1935).
7. Strong symmetrical cut sets of closed euclidean n-space. Fundam. math. 27:136-139 (1936).
8. A characterization of manifold boundaries in E_n dependent only on lower dimensional connectivities of the complement. Amer. math. soc., Bull. 42:436-441 (1936).
9. The sphere in topology. Amer. math. soc. Semicentennial addresses, pp.136-184 (1938)

10. The property S_n. Amer. jour. math. 61:823-832 (1939).

WOJDYSLAWSKI, M.

1.*Sur les rétractes par déformation des coupures de la surface sphérique. Studia math. 9:166-180 (1940)

2. Sur la contractilité des hyperespaces de continus localement connexes. Fundam. math. 30:247-252 (1938).

3.*Rétractes absolus et hyperespaces des continus. Fundam. math. 32:184-192 (1939).

GENERAL BIBLIOGRAPHY

BEGLE, E.

[a] Locally connected spaces and generalized manifolds, Amer. jour. of math. vol. 64 (1942).

BORSUK, KAROL

[a] Über eine Klasse von lokalzusammenhängenden Räumen, Fundam. math. 19:220-242 (1932).

[b] Zur kombinatorischen Eigenschaften der Retrakte. Fundam. math. 21:91-98 (1933).

[c] Un théorème sur les groupes de Betti des ensembles localement connexes en toutes les dimensions $\leq$ n. Fundam. math. 24:311-316 (1935).

[d] Sur le plongement des espaces dans les rétractes absolus. Fundam. math. 27:239-243 (1936).

HUREWICZ, W.

[a] Über Einbettung separabler Räume in gleichdimensionale kompakte Räume. Monatsh. f. Math. u. Phys. 37:199-208 (1930).

[b] Dimensiontheorie und Kartesische Räume. Akad. van wetens., Amsterdam, Proc. 34:399-400 (1931).

[c] Über Abbildungen von allgemeinen topologischen Räumen auf Teilmengen Cartesischen Zahlenräume. Akad. wiss. Wien. Math.-Naturwiss. Kl., Anzeiger 12:97-98 (1931).

[d] Über Abbildung von endlichdimensionalen Räumen auf Teilmengen Cartesischer Räume. Preuss. akad. wiss. Sitz. ber. 1933:754-768.

[e] Homotopie, Homologie und lokaler Zusammenhang. Fundam. math. 25:467-485 (1935).

KURATOWSKI, K.

[a] Sur un théorème fondamental concernant le nerf d'un système d'ensembles. Fundam. math. 20:191-196 (1933).

[b] Sur le prolongement des fonctions continues et les transformations en polytopes. Fundam. math. 24:259-268 (1935).

[c] Sur les espaces localement connexes et péaniens en dimension n. Fundam. math. 24:269-287 (1935).

[d] La notion de connexité locale en topologie. Enseign. math. 35:229-240 (1936).

LEFSCHETZ, S.

[a] On compact spaces. Ann. of math. 32:521-538 (1931).

[b] On separable spaces. Ann. of math. 33:525-537 (1932).

[c] On generalized manifolds. Amer. jour. math. 55: 469-504 (1933).

[d] On locally connected and related sets. Ann. of math. 35: 118-129 (1934).

[e] On the fixed point formula. Ann. of math. 38: 819-822 (1937).

[f] On locally connected sets and retracts. Nat. acad. sci., Proc. 24:392-393 (1938).

[g] On the mapping of abstract spaces on polytopes. Nat. acad. sci., Proc. 25:49-50 (1939).

NÖBELING, G.

[a] Über eine n-dimensionale Universalmenge in R_{2n+1}. Math. ann. 104:71-80 (1930).

PONTRJAGIN, L. and TOLSTOWA, G.

[a] Beweis des Mengerschen Einbettungssatzes. Math. ann. 105:734-745 (1931).

SCHAUDER, J.

[a] Der Fixpunktsatz in Funktionalräumen. Studia math. 2:171-180 (1930).

TUKEY, J. W.

[a] Intrinsic metric of a polytope. Nat. acad. sci., Proc. 25:51 (1939).

WILDER, R. L.

[a] Generalized closed manifolds in n-space. Ann. of math. 35:876-903(1934).

[b] A characterization of manifold boundaries in E_n dependent only on lower dimensional connectivities of the complement. Amer. math. soc., Bull. 42:435-441 (1936).

WILSON, W.

[a] On ϵ-nets in a complex. Compositio math. 4: 287-293 (1937).

INDEX

Ingram Content Group UK Ltd.
Milton Keynes UK
UKHW011243290523
422449UK00001B/65